Control of Fish Quality

Fish is one of the most highly perishable commodities and the public requires continuous reassurance about its quality. Consumer taste is moving to more highly perishable fish products which require more careful attention in preparation and storage. This book provides sound practical advice in as non-technical language as possible on the quality control of fishery products. It was written for fish technologists, industrial technical managers and fishery administrators, indeed everyone in a position of responsibility who needs to know about fish quality and the safety of fish products, also how these can be regulated in promoting improved processing techniques.

The Author
Dr J J Connell is former Director of the Torry Research Station of the UK Ministry of Agriculture, Fisheries and Food and as such is in a unique position to write an authoritative work of this kind.

Control of Fish Quality

J J Connell *CBE*, *BSc*, *PhD*, *FRSE*, *FIFST*

formerly of *Torry Research Station, Aberdeen, Scotland*

Third edition

Fishing News Books

First Edition published by Fishing
   News Books 1975
Second Edition published 1980
Third Edition published by Fishing News
   Books, a division of Blackwell
   Scientific Publications, 1990

British Library Cataloging in Publication Data
Connell, J.J. (John Jeffrey)
   Control of fish quality – 3rd ed
   1. Food: Fish. Quality control
   I. Title
   664.9497

ISBN 0-85238-169-7

Fishing News Books
A division of Blackwell Scientific
   Publications Ltd
Editorial Offices:
Osney Mead, Oxford OX2 0EL
   (Orders: Tel. 0865 240201)
25 John Street, London WC1N 2BL
23 Ainslie Place, Edinburgh EH3 6AJ
3 Cambridge Center, Suite 208, Cambridge
   MA 02142, USA
54 University Street, Carlton, Victoria 3053,
   Australia

Set by Excel Typesetters Company, Hong Kong
Printed and bound in Great Britain by the
University Press, Cambridge

Contents

List of illustrations

Preface

The examination before sale of fish and fish products for evidence of spoilage, damage, adulteration or disease is a long established practice, the object being partly to guard the health and pockets of customers, and partly to ensure the goodwill of vendors. Laws which require that only fresh fish is offered for sale to the public date back to classical antiquity. It is over the past hundred years, however, that official authorities, as part of their general food laws, have taken an ever-increasing interest in protecting the customer against poor quality fish. Many countries now have comprehensive systems of inspection and control of at least some aspects of fish quality. For motives of good commercial practice and profitability the fish industries have also increasingly striven to improve fish quality. Complex preservation techniques like freezing and canning which require for their success close control of operational details and materials have been introduced on an ever widening scale. Fish is one of the most highly perishable commodities and the public has always required continuous reassurance about its quality. Unlike most raw materials fish is not a single commodity but consists of a large number of species of widely differing appearance and flavour. Because of this variety customers are often unsure if particular species or products made from them are good to eat. Thus, from several points of view fish quality has a special importance and there are several reasons why this is likely to increase.

In the first place, the growing sophistication and variety of products and of markets is leading to a greater complexity in the numbers and kinds of quality factors which have to be taken into consideration. Consumer taste is moving away from traditional long-keeping and intrinsically stable products to fresher, milder, more highly perishable ones which require more careful attention in preparation and storage. Consumers are also becoming more aware of possible hazards, malpractices and mistakes arising from food and are individually and collectively becoming more demanding in respect of freshness, naturalness, microbial safety, freedom from pollutants, protection from damage, and convenience. The important benefits that fish make to a health-giving diet are increasingly recognised. On the industrial side, the ever growing size of companies means enhanced financial risks attached to failure to maintain quality and thus a greater need to exercise vigilance. In developed fishing nations there has been a move towards the exploitation of new fish stocks as traditional species become fully or over-exploited. New methods of recovering edible fish flesh are being introduced on a bigger scale. Large companies experienced in food processing and retailing are either taking over smaller fish companies or entering fish marketing for the first time. They are bringing ideas and pressures related to the improvement of quality that are new to many traditional fish companies. There is a growth in brand labelling of fish products. All these developments have created new quality problems. On the other hand, developing countries are building up capacity to exploit their fish resources, are striving to reduce waste caused by spoilage and are increasingly entering export markets for fish products – all trends which create demands for more quality control and inspection. Finally, the effect of international collaboration in the formulation of quality standards for fish and fishery products and the adoption of these standards by individual countries and organisations must be noted.

For all these reasons it seems useful to draw together in one text those topics which deal with fish and fishery product quality and the measures which can be taken to regulate it or to control adverse changes in it. A number of books in different languages cover some of these topics individually or in a very specialised way and I acknowledge my indebtedness to their authors.

Here the aim is to present a balanced, discursive and coherent account of the subject couched in as non-technical language as possible. However, since we are dealing often with technicalities, some special terms and concepts must be used, and will be explained where necessary. Also, some familiarity with fish, fishery products and processes is assumed: the book is concerned primarily with principles and is not a treatise on all aspects of handling, processing or treating fish. It does not claim to present any radically new proposals or theories, but the hope is that it will provide sound, practical advice for those who are confused by the quantity of sometimes conflicting information on the subject or who are without advisers. It is written for fish technologists starting out in this field, industrial technical managers and fishery administrators; indeed, all those in positions of responsibility who need to know more about fish quality and the safety of fish products and how these can be regulated. The intention is to draw together succinctly information which is at present scattered in many places and to lead the reader to where further useful details can be obtained. Rather than interpolate detailed references I have chosen to list at the end publications which I consider give an appropriate and reasonably comprehensive account of particular topics. The illustrations are specific to fish: equipment and methods common to food industries generally are not included. Latin names for fishes are included where I judge some ambiguity might arise.

In the space available it is not possible to include separate discussions of the very numerous species and individual products marketed throughout the world. Therefore, the approach adopted in the main is to group together species or important types of product that present similar problems of quality control. The following special products have been omitted altogether: stabilised, special purpose frozen fish minces such as the Japanese product 'surimi'; Japanese 'kamaboko'; fish sausages; fish meals and oils; pet foods and special animal feeds; silage; fish protein concentrates; fish pastes and spreads; and fermented products in which enzymes are deliberately used to enhance the flavour and texture of the natural fish. Unless it is necessary to make the distinction, fish will be taken to include shellfish. The term 'fish industries' as used in this book means all those enterprises engaged in conveying in all its forms fish from the waters in which it lives to the final consumer.

The opening chapter develops the basic question of what is fish quality. The second chapter deals with the manifold aspects of intrinsic quality. Chapters 3, 4 and 5 discuss the various ways in which quality changes in fish and fish products, the factors that affect these changes and how they can be brought under control. Chapter 6 describes methods for assessing quality. The remaining chapters discuss the organisational and systematic means available to help the problem of getting to the consumer the best and safest fish at an acceptable price.

I have to thank my colleagues, Mr Peter Howgate and Dr Geoffrey Hobbs, for checking parts of the text. A previous Director of Torry Research Station, Dr Geoffrey Burgess, read the whole draft and corrected many errors of fact and infelicities of style; my debt to him is inestimable. Naturally, I take sole responsibility for any remaining faults in the book.

Figure 2.6 was provided by the Burnham-on-Crouch laboratory of the Ministry of Agriculture, Fisheries and Food. Figures 3.2–3.7 are taken from the Advisory Note Series of Torry Research Station.

1 What is quality?

Before embarking on the subject of control it is necessary to be clear about what is being controlled. Quality is defined in dictionaries and glossaries in simple phrases such as: the totality of features and characteristics of a product or service that bear on its ability to satisfy a given need; degree or grade of excellence; grade of goodness. However, these definitions do not take us forward very far and the full range of meaning of the term quality as applied to a particular commodity can only come from a wealth of practical examples. The next four chapters provide such examples for the range of fish products covered in this book. The meaning of quality adopted is deliberately wide, namely all those attributes which consciously or unconsciously the fish eater or buyer considers should be present. Thus, quality will embrace intrinsic composition, degree of contamination with undesirable materials, nutritive value, degree of spoilage, damage, deterioration during processing, storage, distribution, sale and presentation to the consumer, hazards to health, satisfaction on buying and eating, aesthetic considerations, yield and profitability to the producer and middlemen. All of these have to be taken into consideration when designing control or inspection procedures or when changing or improving systems of handling and processing.

Fish is meant to be eaten and decisions about what constitutes gastronomic quality rest ultimately with the consumer, that is the person who eats the fish. Thus, in order to discern those attributes that require controlling and the particular values they should assume the opinions of consumers must be elicited. In matters of ordinary sensory appeal this is done either through the techniques of market research or more usually by building up knowledge of those attributes that lead to profitable sales and those that do not. Consumers' attitudes to fish quality are not immutable and it is necessary to be informed of changes in these attitudes as they occur. It is suspected that in-bred traditional views of what consumers will accept sometimes have acted as impediments to the fully rational exploitation of fish resources. The importance to the development of a valid view of quality of an up-to-date awareness of market requirements cannot be overstressed. The producer who wishes to know what aspects of quality are important will be in close liaison with his customers, will react quickly to their changes in taste and will keep systematic records of their reactions to, or complaints about, his products. It is equally necessary to be aware of the economic factors that affect attitudes to quality, such as cost, supply and demand. Some improvements in quality or the achievement of consistent quality can be effected at no extra cost, but some cannot. Extra costs must be balanced against the long-term benefits of continued customer loyalty and of increased sales.

For those attributes that affect health and safety the ordinary consumer cannot be expected to be able to have a fully informed opinion. In these circumstances the Government or other independent official body has to step in to act on behalf of consumers and by, for example, the enforcement of standards or drawing up of regulations, safeguard their interests. Officials need, therefore, to have an up-to-date knowledge of such matters as composition and possible sources of hazard to health.

In the course of reaching the consumer, fish is often involved in several commercial transactions and at each of these a knowledge of the standard of quality expected by the consumer must be transferred back until it eventually reaches the primary producer. Consequently, the primary producer's idea of what is acceptable to the consumer may be rather different to what the consumer himself thinks. Also, quality often inevitably falls during handling, processing and distribution. For example, in the case of the freshness of chilled fish the standard of quality will usually have to be higher at the first exchange (e.g. port market auction) than at the consumer end. The same is true of the quality loss of frozen fish.

Processing may impose its own independent and additional quality standards. For example, in order to attain the most highly profitable yield of product a processor may require from his supplier raw material of a particular size range. This range is then the best quality standard for this particular purpose.

Although a good deal undoubtedly exists, little published information is available that describes consumers' attitudes to fish quality, or indeed food quality generally. This is understandable because such information can be of commercial value and is confidential to customers. However, in recent years numbers of official national and international quality and grading standards have been published which provide for some products good indications of what is minimally acceptable. Insofar as these standards have been drawn up in collaboration with industry and other interested parties, it can be assumed that they are realistic reflections of what consumers want or will accept. It is perhaps surprising that it has been possible with relative ease to reconcile widely different consumer tastes in single product standards of world-wide scope.

Quality control has been defined in a number of ways. Two examples are:

(i) the operational techniques and activities that sustain the product or service quality to specified requirements, and the use of such techniques and activities.
(ii) an industrial management tool designed to ensure maintenance of quality at a level that satisfies the customer and that is economical to the producer and seller.

Again, the full implications of these words can only be revealed by the many kinds of practical examples given in what follows.

2　Intrinsic quality

By intrinsic quality is meant the sum of attributes that are inherent in the harvested raw material. By far the largest part of the fish used for food in the world is caught in the wild and for this part fully active control over or manipulation of intrinsic quality is impossible. In this respect fish industries are at a major disadvantage *vis-à-vis* food industries handling cultivated and therefore inherently more controllable commodities. Nevertheless, two types of control are possible; both are inefficient but are used extensively. The first is conscious selection of grounds, seasons, or fishing methods which are likely to result in either fish of the desired quality or the avoidance of fish of undesirable quality. The second is selection or sorting from mixed catches, of fish into desired qualities.

In the case of the relatively small quantity of husbanded or cultivated fish more continuously active control of intrinsic quality is feasible since all aspects of the primary production can be in principle manipulated. Consequently, it becomes possible more or less at will to produce fish of a certain desired kind, size, composition and keeping quality.

The factors that affect intrinsic quality will be considered under the eight headings that follow.

1 Species or identity

In all communities some species are more highly valued than others and these preferences are very stable through time and different generations. A full description of national and regional tastes in fish is out of place here and it is sufficient to make a few illustrative observations. Such preferences are largely but certainly not entirely a consequence of the particular range of species traditionally caught and presented to consumers. 'We like what we know' and what we know is that which has been served to us throughout upbringing and taught as being desirable. In the United Kingdom for example the range of species eaten is relatively narrow and has hardly changed since the time centuries ago when the pattern of catching was established by the species accessible to the first inshore fishermen. On the other hand the Japanese, with more adventurous early fishermen, waters harbouring a wider range of species and difficulties of supplying sufficient protein from the land, have established a much wider catholicity of taste.

Other influences on preferences include the objectionable external or flesh appearance of some species, taboos and the presence in the edible portion of excessive numbers of unpalatable bones. In some cases beauty of appearance or taste is held in high esteem. The pleasurable feeling engendered in many people when eating, say, lobster results in this species receiving high marks for quality. The deliciousness of Japanese fugu overcomes the real risks (to be described later) involved in eating it. The brilliant but fugitive coloration of some snappers makes them an ideal showpiece in a ceremonial meal; they are correspondingly expensive. Degree of availability can influence what people consider to be quality. Thus, species like halibut or salmon which nowadays are ranked as high quality species were in certain communities so plentiful as to be relegated as very poor eating or as worthless. Otherwise, it is difficult to rationalise. Why, for example, do the British prize so little saithe, redfish (*Sebastes* spp.) mackerel and squid which are relished in some other European countries? The low esteem in which these so-called 'rough' species are held is not helped, of course, by the practice among fishermen and the trade generally of handling them in such an inferior manner that their quality suffers unnecessarily.

The control of harvested fish species involves no more than careful selection on the basis of a knowledge of what is marketable. Pressure to meet specific market requirements for individually sorted species may be tempered in the future by the continued growth of manufactured products such as Japanese surimi (minced, frozen flesh) and kamaboko, coated fish fingers and portions, in which the exact character of the species used is of reduced importance.

Although not directly related to market value it should be noted that the rate of spoilage or deterioration is species dependent. When chilled or frozen, fatty fish like sardines and herring spoil or deteriorate more quickly than lean fish. Round lean fish like cod and pollack similarly tend to be poorer in this respect than flat fish. Fish caught in warmer waters tends to keep better during chilled storage. Some of these points will be developed later.

Although publications on quality often present illustrations of food fishes, no attempt is made here to follow the practice. Such a compendium would outrun the intended scope and be somewhat redundant. The references at the end will, it is hoped, serve to fill the omission.

2 Size

In general, large fish of a given species fetch the highest prices. Consumers are prepared to pay more for large examples of commodities like shrimps, crab, lobster and cod or portions cut from large halibut and salmon because they are visually and gastronomically more satisfying. Nevertheless, there is no evidence that size (which within a species is related to age) is in any way related within one species to flavour quality. That is to say large fish are not necessarily more finely flavoured or textured than small. Being able

Fig. 2.1 Fish of uniform size laid out at an auction market; such sorting is achieved by eye and hand both on board and on unloading from the fishing vessel: the labels, which denote size and freshness, usefully contribute to the control of quality.

to cut off and masticate large pieces of flesh can be more pleasing than dealing with morsels. Processors often place a high value on large fish because percentage yield of edible material is high, handling costs per unit weight are lower, they often keep better and often more uniform products can be made from them.

On the other hand, for some purposes the optimum size is less than the largest. Large sizes of trout, clams and oysters are not favoured for table use because portion size is then too big or expensive. For canning, specific sizes of sprat, herring, sardine and similar species are required to ensure correct can-fill. Cutting or filleting machines can usually only be adjusted to accept a limited size range of fish. For the machine to give maximum yield or even to work at all, it is necessary to segregate mixed batches of fish into particular size ranges suitable for particular machine settings.

Control of size is effected in the first instance by choosing those fishing grounds, seasons or methods which are more likely to yield marketable selections. Clearly, this is a somewhat hit and miss method guided by largely undocumented fishermen's lore and experience. Some preliminary sorting of the catch for size is often carried out on the fishing vessel. Almost always this has to be effected manually: in the current state of development, machines for size grading are too bulky to be used on any but the largest

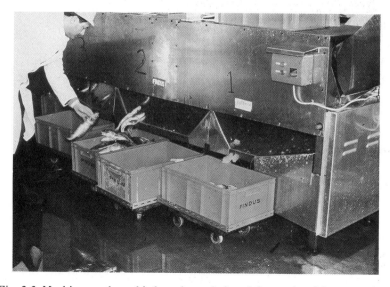

Fig. 2.2 Machines such as this have been designed for sorting fish automatically into different size ranges.

fishing or factory vessels. At later stages selection or grading can be done visually and manually at port markets or mechanically in the processing factory. Mechanical grading is also carried out at some port markets. Cultivated fish are often grown to specified limits or the stock culled. There is a growing tendency at port markets to size grade fish more thoroughly and in a more standard fashion.

3 Sex

For some purposes males of a given species are valued differently from females. The excised roe of female fish may be valued as a commodity in its own right and such fish prized on this account alone. The canned product made from female capelin (*Mallotus villosus*) is very different from that made from male capelin.

It is rarely possible to fish with the specific intention of catching only fish of one sex and it is almost always necessary to depend on the chance occurrence of the particular sex required. As mentioned in Chapter 6, segregation of single sex fish from a mixed batch is in some circumstances possible to achieve mechanically.

4 'Condition' and composition

It has long been known that in all species of fish, seasonal changes in certain bodily characteristics occur. At certain times ordinary fish appear thinner, flabbier and less lively than at others, the flesh being more watery and softer and containing less protein and fat. Fish of this kind is said to be in poor 'condition' or 'out of season'; it has poor sales appeal and gives lower yields.

Poor condition occurs at the period after the fish have spawned. For instance, in the case of many species in temperate or Arctic waters this is the spring. Just before spawning and during it, food reserves in the flesh, and in some species in the liver, are transferred for the development of the gonads (eggs and spawn). During spawning and for some period afterwards most fish do not feed (elasmobranchs, that is sharks, rays, etc., are exceptions). As a consequence of both effects the flesh after spawning becomes severely depleted of protein, carbohydrate and fat and the fish are accordingly 'run down'. Similar poor condition can arise, however, when for whatever reason the fish are not feeding or are feeding at

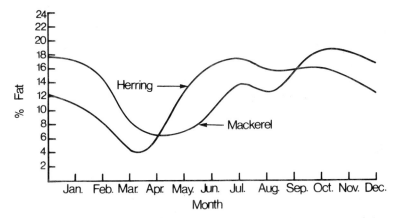

Fig. 2.3 The fat content of pelagic fish can vary considerably throughout the year; the graph shows the average fat contents of batches of herring landed at the same port in W. Scotland, and of batches of mackerel landed at the same port in S.W. England.

an abnormally low level. Once fish start feeding again they normally recover their good condition. Seasonal, cyclical changes in flesh composition are observed in all species though less noticeably in some shellfish where changes in glycogen content are a particular feature.

In white fish in poor condition the flesh when cooked and eaten is soft, gelatinous or sloppy; whether the flavour is weak or insipid as is sometimes alleged is a matter of some uncertainty but in any case it is not an overriding factor in marking down the quality. In occasional extreme cases observed in halibut, rockfish (catfish; *Anarhicas* species), large roughback (American plaice; *Hippoglossoides platessoides*) and some others, depletion of the protein gives rise to a jelly-like, completely unusable, state in the flesh. This condition should be distinguished from a similar one which is an intrinsic characteristic of species like the smoothhead (*Alepocephalus bairdi*). In this fish the normal water content is very high (93%) and the protein content is very low (6–7%) which gives an unpleasant junket-like and watery texture to the cooked flesh.

It is in the numerous fatty, pelagic species like sardines, sprat, herrings, mackerels and anchoveta that compositional and dependent quality changes are most marked. The fat content of herring flesh, for example, can change from below 1% to over 25% between the starvation period after spawning and the height of the feeding period. During this change overall weight is maintained almost constant by a corresponding reduction in water content. For processes like canning and kippering which utilise these species a high fat content is highly desirable if the best quality products with excellent appearance, succulence and flavour are to be achieved. When meant to be eaten raw as sashimi, tuna of high fat content is of much greater value than of low. Fish of all those species in the post-spawning, lean and so-called 'spent' condition make poor-quality articles. For some special products, for example canned brisling, the fat content of the sprat raw material should not exceed 12–13%; fish of higher content result in unsightly oozing of fat in the can. During intensive feeding periods the fat in pelagic species is rapidly laid down in very fluid layers just under the skin; fish in this condition are difficult to brine successfully. Similarly, pelagic fish (and indeed other fish) after heavy feeding ('feedy' fish) are susceptible to the damaging condition known as 'belly-burst' which will be explained in the next chapter.

Further differences in intrinsic composition within one species can give rise to complicating secondary influences on quality. Lean fish in poor condition spoil when chilled more rapidly than the same species in good condition. The reason is probably connected with the higher flesh pH of the former which arises primarily because the glycogen content is lower. After death the glycogen is converted into lactic acid, the concentration of which determines the flesh pH; lower lactic acid concentration results in higher flesh pH. Bacteria which cause spoilage are more active in flesh of higher pH. Winter flounder (*Pseudopleuronectes americanus*), a relatively lean fish, goes rancid during frozen storage more rapidly when the fat content is at its highest average value in summer months. Cod in poor condition with high flesh pH 'gape' less than those in good condition with low pH. 'Gaping' is the severely damaging tendency of fillets, particularly those cut from thawed whole frozen fish, to split into fissures and holes.

It has been noticed in the United Kingdom that at certain times of the year difficulty is experienced in making by a standard procedure void-free blocks of frozen cod fillets. Good quality blocks should be as void-free as possible. These difficulties are associated with an as yet unidentified intrinsic and seasonally-occurring physical or biochemical property of the fish. Commercial losses occur through the occasional incidence in halibut of the condition known as 'chalkiness'. The raw flesh cut from fish stored for some time in ice appears white and dull as if it had been cooked. This condition occurs in fish whose flesh pH falls below a value of about 6 after

Fig. 2.4 One cod fillet exhibits an extreme degree of 'gaping', and the other none.

death. It is inferred that such fish contained in life a high glycogen content brought about by especially heavy feeding. The final pH of cod flesh also influences the rate at which deterioration occurs during frozen storage; fish of low pH tend to toughen more rapidly.

'Honeycombing', the descriptive textural defect of tuna, particularly the canned article, is similarly probably associated with the small proportion of fish having exceptionally low pH, though there is also some evidence of an association with rather advanced spoilage. In some species the protein collagen content of the flesh and skin tends to increase as the fat content falls. This has the effect of strengthening the tissue so that the fish hold together better when heat processed, as in canning. Taking these examples together it can be seen that although fish in 'good condition' may be most suitable for some purposes, for others involving further processing it may be the worst possible choice.

The exact time of year when spawning and poor condition occur varies with fishing ground and in some fisheries may extend over many months. Sometimes pelagic species when caught at the same time and place will contain groups spawning at two or more distinct times of the year. In addition, fish landed from vessels which visit several grounds may be in various states of condition. This, coupled with the advent of freezing and cold storage, means that as far as the user and consumer of fish is concerned the old concept of fixed seasons when fish was in prime condition has lost a good deal of its general validity.

Good control of intrinsic quality of this kind is possible by judicious avoidance of particular fishing grounds and seasons known from experience to yield fish of poor condition or poor potential for processing. In Norway a statutory seasonal limitation is placed on the sprat fishery in order to ensure that raw material of poor quality is not used for canning. In those situations where batches of varying condition or composition are available it may be worthwhile, assuming it is possible, to select visually those fish which are suitable for particular purposes. Means for automatically sorting mixed batches of this kind are not available. It is now quite common for larger processors or distributors to specify the fat content of raw material or products. The supplier must then measure the average fat content of the batches that he has available. A method described in Chapter 6 is suitable for this purpose.

5 Parasites and other organisms

A parasite is an organism living on or inside another and depending upon it for some of its vital needs, particularly nutriment. Fish, in company with all other animals, are liable to be infected by parasites some of which have complex life histories. Although the fish processor can soon learn to recognise the commoner types which may affect the major species of fish he handles, he cannot be expected to be an expert in this difficult field.

It is impossible here to do more than mention the major types of fish parasite and how they may be recognised. Fortunately, although many parasites are unsightly and can therefore give rise to complaints if they are discovered by the consumer, few are harmful. Furthermore, many live on the surface of the fish or in parts, such as the viscera or the head, that are not eaten.

Thorough cooking kills all parasites and so renders them completely harmless. Those that can cause serious diseases in man do so mostly in those parts of the world where fish and shell fish are eaten raw. Mention of these conditions will be made later; here it is stressed that those concerned with quality control must obtain specialist advice in those countries where the risk of such diseases is known to exist, and must follow strictly the procedures that are laid down by experts in this field.

Parasites and the conditions they cause in fish are not always easily identified; even within one group, for example the roundworms, their effects can be extraordinarily varied and exact identification of the species of parasite involved is usually a job for the specialist. Nevertheless, the quality controller should be sufficiently aware of the commoner manifestations of parasitic infections to be on the lookout for them as a matter of routine and to be able to recognise them at a glance.

There is great diversity in the various types of parasites; those in fish mostly belong to the following groups of organisms: Protozoa, flatworms, roundworms and crustacea. Micro-organisms called fungi and bacteria also cause disfiguring diseases in fish and will be mentioned. What follows is not, however, an exhaustive scientific account but rather is intended as a guide to the practical man in industry.

(i) Protozoa

These are microscopic, often one-celled, organisms. Most are free living and harmless but a few cause disease in man and animals. Protozoal infections in fish of importance to the processor generally occur in the flesh, where they may cause severe localised or generalised softening. Some species may cause lesions on the skin and these forms may be particularly important in fish farming.

A well-known protozoal infection is that caused by the myx-osporidian *Chloromyscium thyrsites*; it occurs in a number of species of fish including Atlantic hake where its presence is difficult to detect in the flesh immediately after capture. After the fish has been stored in ice for a few days, however, the flesh of badly infected specimens becomes softened by proteolytic enzymes produced by the parasite and appears like toothpaste. It is well named 'milky hake'. Other species of fish affected by this or related organisms include snoek (*Thyrsites atun*), John Dory (*Zeus faber*) and swordfish (*Xiphias gladius*). Salmon and halibut are also sometimes seriously infested by protozoa called *Henneguya salminicola* and *Unicapsula muscularis*, respectively. In some species the flesh is pitted by ulcerated spots full of white pustular material.

(ii) Flatworms (platyhelminths)

Many flatworms are free living but the flukes, or trematodes, and tapeworms, or cestodes, are all parasites. Most flukes and all tapeworms have complex life histories involving stages in different animals in addition to fish. Animals harbouring these stages are referred to as intermediate hosts. The flatworms found in fish are almost all harmless to man.

Adult fish tapeworms always occur in the gut and do not therefore present a problem. Adult flukes likewise usually occur in the viscera. Larval tapeworms may, however, be found in the flesh of many fish, for example halibut, weakfish (*Macrodon ancylodon*) and some whitefishes (for example, *Coregonus clupeaformis*). One of the most common trematodes is the very small and harmless larva of *Cryptocotyle lingua* which exists in a dark-coloured cyst on or just under the skin of gadoids, herring, mackerel and other species. It gives rise to a typical peppered effect of dark spots which is re-moved by skinning.

Two important diseases in man are caused by eating live cysts of two special types of flatworm. These diseases occur in countries where fish is eaten raw or only partially cooked.

The lung fluke (*Paragonimus*) occurs in man and animals in Asian countries. Intermediate hosts are freshwater snails and crustacea

Fig. 2.5 The peppered appearance of the eye and disfigurement of the lips of this fish are due to the presence of larval trematodes.

such as crayfish and freshwater crab. Man can become infected by eating 'raw' infected crustacea; the adult fluke, which is only a few millimetres in length, lives in the lung. Very large numbers may be present in severe cases.

The broad tapeworm of man, *Diphyllobothrium latum*, was formerly common in some European countries. When present in man it inhabits the intestine. Fish capable of acting as intermediate hosts commonly include salmon, trout, pike and perch. The disease is nowadays rare since the risks of eating raw fish of these species are well understood.

(iii) Roundworms (nematodes)

Nematodes may be found in the gut and viscera of fish or encysted in the flesh. It is the encysted forms that generally cause the problems for the processor since it is difficult to remove them all and they are often unsightly. The cod worm, *Porrocaecum (Terranova) decipiens*, is a well-known example of a nematode parasite; as the brownish-coloured larval form it exists coiled up in cysts in the flesh. The adult occurs in the gut of seals and the level of incidence of the worm in the flesh of particular stocks of cod is related in part to the level in the nearby seal population.

Fish nematodes do not cause recognised parasitic diseases in man but the larvae of one nematode called *Anisakis* larvae can cause serious inflammation of the stomach or gut wall, if eaten live. Man is not the natural host of this parasite and thus the attack on him is purely adventitious. Cases, which nowadays are very rare, have been traced to herring in Europe and squid in Japan. Nematodes cannot survive prolonged freezing and cold storage; Norwegian, Dutch and Japanese regulations therefore require certain types of fish product which are eaten raw to be frozen and stored at −20°C for at least 24 hours before being released for sale. Nematodes are not readily killed by modern methods of curing involving only moderate concentrations of salt.

(iv) Crustacea

Crustacea contain many parasitic forms, most belonging to the sub-group known as copepods. Although some are clearly recognisable as Crustacea others have become so highly parasitic that their affinities are not immediately recognisable. The sea louse, *Lepeophtheirus*, which is to be found on fresh-run salmon and sea trout, is an obvious crustacean. On the other hand, *Sphyrion lumpi*, which occurs in the flesh of redfish, *Sebastes marinus*, is quite unlike free-living Crustacea. The external part of the organism, which is almost entirely reproductive in function, is attached by a slender stalk which passes through the skin of the fish to a dumb-bell shaped head embedded in the flesh. The external part of the parasite is not infrequently detached by the time the fish reaches the production line and all that remains is therefore a hard brown or black mass embedded in the tissue.

(v) Fungi and bacteria

Fungal infections are a well-known hazard for the tropical fish fancier; they are in fact widespread, though few are of importance for the fish processor. They are often only distinguished from the bacterial and protozoal infections by the expert using a microscope.

An example of a fungal disease is the condition known in the UK as 'greasy haddock'. It is caused by *Ichthyosporidium hoferi*. The flesh of infected specimens is soft and has a somewhat sweetish, slightly sickly smell. When the fish is smoked white spots are to be seen in the flesh. The condition tends to be localised to certain fishing grounds and also appears to be seasonal.

Bacterial infections are also widespread and can be of particular importance in fish farming. Lesions, nodules and pustula areas found in commercially caught fish may all be attributable to bacteria. It is not possible here, however, to describe these in detail and in practice the fish processor will find such conditions to be relatively rare and more a matter of passing interest than of commercial importance.

None of these disfiguring micro-organisms is harmful to man, that is, none is pathogenic. However, a few other micro-organisms of different kinds are pathogenic: they will be described later.

(vi) Control measures

Fundamental control of parasites in nature, that is prevention of entry into fish, is virtually impossible. In some instances partial control might be feasible by reducing the numbers of previous hosts, for example seals in the case of 'cod worm' parasite. In small bodies of water more effective control is possible by proper husbandry and hygiene. For the most part control has to be by avoidance of fishing areas known to yield heavily infested fish or by selection of the catch. If parasites are visible on the whole fish they can sometimes be removed; otherwise the catch has to be discarded. Official quality or public health inspectors will often condemn as unfit for human consumption visibly parasitised fish. Visible parasites inhabiting the flesh will not become apparent until the fish are filleted or cut up, though clues about this kind of incidence sometimes can be obtained by an examination of the viscera or body cavity. Filleting or handling operatives are usually given instructions or are knowledgeable enough to discard or trim flesh showing a sufficiently high visible surface incidence. The presence of parasites embedded in relatively thick parts often can be detected by the process known as candling which consists of placing the skinned fillet or piece of flesh flat on a translucent surface illuminated from below. Given the right conditions parasites can be excised by forceps or a suitable bent piece of wire. Candling and excision add considerably to handling costs and are only resorted to with valuable fish, low labour costs or stringent customer requirements. Sometimes, as is evident from the Appendices, customers specify maximum allowable incidence, this being expressed in terms of the number of easily detectable or objectionable parasites per unit weight or surface area.

6 Naturally toxic fish

The vast majority of food fishes are intrinsically safe to eat. Unfortunately some species are naturally toxic and cause injury to health or even death when through accident or ignorance they are eaten. The majority of the species involved are caught in tropical or sub-tropical areas and it is only here that general systematic control measures are considered desirable. Nevertheless it is as well for anyone with responsibility for handling or importing fish in quantity to be aware of the problem. Poisonous fish of this kind are referred to as biotoxic (naturally toxic) to distinguish them from fish that become poisonous or harmful through contamination with chemicals or organisms resulting from man's pollution. Some are venomous, that is capable of inflicting injury through stings, bites or cuts, but we are not concerned with these here.

There are three main types of fish poisoning: ciguatera, puffer (globefish or fugu in Japanese) and paralytic shellfish poisoning.

The name for the first is derived from the trivial Spanish name for a mollusc (*Turbo pica*). It is however caused by eating the flesh of a wide variety (one book lists several hundred species) of mainly carnivorous fish and shellfish inhabiting shallow waters in or near tropical and sub-tropical coral reefs. The symptoms, rarely fatal, include diarrhoea, nausea, vomiting and tingling sensations in the extremities. Cooking does not destroy the toxic principle. Fish of the incriminated species are not invariably toxic and the reasons why they suddenly become so and as quickly become non-toxic again are obscure but are probably connected with changes in the nature of their food. For this reason quality control measures are difficult to apply. The most than can be done is to avoid marketing species that have a known consistent record of toxicity. Here it is necessary to fall back on a well-informed and alert trade and, where it exists, the local fishery or health inspectorate. As an aid in this direction a guide book on poisonous fishes has been published in Japan for distribution to the trade. At Tokyo fish market suspected fish are removed officially and tested on cats and mice for toxicity before release. Some years ago in Italy several unfortunate people died, probably as a result of eating this type of toxic fish imported by an unwary trade.

Poisoning by eating puffer fish (*Tetraodontidae*) is much more serious in that the mortality rate is over 50% but on the other hand these fish are only used for food in a few countries, notably Japan, where, despite all precautions, several persons a year die from this affliction. The flesh is non-toxic or only slightly toxic but the viscera in particular is extremely dangerous and the risk normally arises from contamination of the flesh with viscera. More rarely the viscera itself is eaten. In Japan only certain persons are allowed to prepare the fish for public consumption. Further than this the remedy would seem to be more education about the risk involved.

Paralytic shellfish poisoning occurs when certain molluscs and in particular mussels and clams, are eaten. It has been noted that these molluscs become toxic only during periods when high concentrations of certain types of unicellular organism, known as dinoflagellates, occur in the sea. These types, which are themselves toxic to man, tend to be pigmented particularly with red colours and in consequence the sea areas harbouring them in high concentrations assume a reddish tinge, commonly referred to as 'red-tide'. Redtides usually occur at warmer times of the year. Molluscs feed on the dinoflagellates and absorb the toxic principle (toxin) without themselves being affected. The toxin, which is not completely destroyed by cooking or canning, gives rise to symptoms of tingling and numbness in the extremities and of muscle weakness which in mild cases passes off quickly. A small proportion of cases prove fatal. The disease has been reported widely throughout the world. Prevention lies in avoiding marketing suspected produce, education and warning of the public against gathering or eating samples from waters known to be affected and the closing of affected areas to commercial harvesting until samples are proved by animal testing to be safe. Once infected the molluscs cannot be readily cleaned of toxin by putting them in pure water. The risks involved in this disease are very small and do not warrant the continuous monitoring for toxicity of the generality of mollusc stocks throughout a country. However, in those isolated areas known to be regularly affected by the condition, it is a wise public health safeguard to monitor the stock routinely. This is done, for example, in the USA.

A few other rarer, authenticated poisonous fish occur; in particular the descriptively-named castor oil fishes (*Ruvettus pretiosus* and *Lepidocybium flavobrunneum*) which have a pronounced purgative effect should be noted.

7 Contamination with pollutants

Seas, rivers and lakes are the eventual sinks for many of the harmful or waste substances disposed by man. Aquatic life, including food fishes, is capable of absorbing and concentrating these pollutants. In some isolated and very rare instances severe injury to health or death have occurred from eating such contaminated fish. Some harmful substances occur, naturally and inevitably, but in 'background' concentrations so very low as to present no conceivable hazard to man. On the other hand, man-made pollution of the marine and freshwater environments has in recent years increased considerably both in extent and in respect of the number of substances involved. There is therefore growing concern throughout the world with the impact of pollutants on the quality and safety of the food man wins from aquatic sources.

As far as effects on the eating quality of fish are concerned the most effective control measures are obviously those that reduce or eliminate pollution at source. To a large extent these are matters of a relatively long-term nature to be tackled at national and international level. Some action of this kind has already been taken and continues to be taken and the results are already showing themselves in reduced or static concentrations of some polluting substances in fish. What is more, several Governments have taken steps to institute programmes for monitoring the concentrations of potentially harmful substances in the aquatic environment, in many cases with particular reference to fish. In some cases experience of monitoring extends over several years. These programmes offer a considerable assurance locally that fish quality with respect to factors of this kind is being safeguarded. On the basis of surveys for pollution a number of Governments have given advice to the public and industry on the safety of fish. In most cases this has been completely reassuring but in others it has been necessary to issue cautionary warnings. For example, the Japanese Government has advised that the concentrations of some pollutants – mercury, cadmium and polychlorinated biphenyls – were sufficiently high in the generality of fish caught in the vicinity of Japan that a very high consumption of this food should be avoided, or for some minor classes of consumers, consumption should be avoided altogether. Other examples of specific advice are given later.

All this does not necessarily absolve the fish industries from an understanding of the issues involved and in some cases from taking steps to carry out their own monitoring programmes or checks. A few Governments have set maximum permitted concentration levels for some pollutants in fish; if these are exceeded the fish is not allowed to be eaten within the country. Clearly, firms exporting or importing large quantities of fish which might become subject to these restrictions would be advised to check consignments in order to safeguard themselves against financial loss.

The main classes of pollutants which need to be considered in some detail are metals and other elements, chlorinated hydrocarbons, mineral oils, radioactive isotopes and micro-organisms.

(i) Metals and elements

A large number of potentially harmful metals and elements are known pollutants but so far only mercury has been implicated in disease to man caused by eating fish. However this reassuring picture should not divert attention from the possibility that, given the right conditions, pollution from any of the substances in this class might cause unsuspected hazards to man. In the main, widespread routine monitoring of fish for these substances is not called for, but spot checks on specific elements may be necessary on fish caught in suspect areas, for example near effluent discharges or waste dumps. The elements of most concern are cumulative poisons, that is those that cause injury to health through progressive and irreversible accumulation in the body as a result of ingestion of repeated small amounts. These include mercury, cadmium, lead, selenium and arsenic.

Between 1953 and 1961 a mysterious disease affecting the central nervous system occurred in persons living in the region of Minamata Bay, Japan. This was ultimately shown to be caused by eating marine fish contaminated with high concentrations of mercury. A similar outbreak occurred in Niigata, Japan in 1964–65. In all, several hundred persons were affected, many extremely severely and permanently; about 50 died. In both cases the cause was industrial effluent containing mercury that entered the sea.

These disasters prompted authorities and analysts in other countries to investigate fish caught in areas possibly contaminated with mercury. At the same time medical experts on the basis of somewhat fragmentary evidence relating to the effects of ingesting different amounts of mercury, calculated provisionally that for a sufficient margin of safety and a fairly high level of weekly fish consumption (four to eight meals of 150g each) the concentration of mercury in the edible portion should not exceed 0.5–1.0 mg/kg wet weight (mg/kg is equivalent to parts per million). In fact, most of the mercury in fish (and in most tissues) is in the form of the compound methyl mercury which is more toxic than mercury itself or its salts.

The food regulatory and health authorities in some countries have taken such a serious view of the hazard presented by possible mercury poisoning that they have issued general proscriptions against the use for human consumption of fish containing more mercury than the values just quoted. Thus, Canada does not permit a level higher than 0.5, whilst the limit in the USA, Japan, Sweden and Finland is 1.0 with the additional proviso that fish with a level of 0.5–1.0 should not be eaten more than about once per week. In the UK no statutory limit on mercury in fish caught and eaten there is considered necessary, though imported fish banned in the country of origin because of high mercury content may not be sold for human consumption. In Italy imported fish must be accompanied by a certificate stating that the consignment does not contain unacceptable levels of mercury.

Analyses of fish flesh (and other foods) in Sweden, Finland, USA, Canada, UK, Japan and elsewhere have revealed that the mercury content of marine fish, although generally higher than most foods, does not for the most part exceed a value of 0.5. Two notable exceptions are the large predatory species tuna and swordfish. When exposed to a mercury-containing diet, fish gradually accumulate the metal throughout life. The higher members of a food chain consisting of fish eaters will tend to contain more mercury than the lower members. Accordingly, most small or medium-sized tuna contain levels that fall below the limit just quoted. On the other hand many large tuna and almost all swordfish of commercial size exceed the limit. As a result of these findings 15–20 years ago much processed tuna and swordfish of high mercury content was withdrawn from sale, the processing of swordfish ceased and careful checks by the canning industry and

Government agencies on the mercury content of tuna raw material instituted. It is uncertain whether the high mercury concentrations in tuna and swordfish are the result of industrial pollution of the high seas. Museum samples of these and other species of fish up to 90 years old contain similar high concentrations. In some cases marine fish caught inshore have rather high mercury contents which often can be associated with local industrial pollution or the occurrence of mercurial ores.

High local concentrations of mercury in the environment occur more often in rivers and lakes; consequently a larger number of freshwater species have been found to contain high levels of mercury. In particular, large quantities of mercury-containing compounds have been discharged by industrial concerns into rivers and lakes in Sweden, Finland and Canada. In these countries considerable quantities of commercially used freshwater fish were affected. Because of the persistence of the pollutant in the food web and in natural sediments, a serious chronic problem now exists in some areas. Sweden and Canada have placed an indefinite ban on fishing in certain rivers and lakes known to be contaminated with high levels of mercury.

The position with regard to the effect on fish quality of cadmium, lead and selenium and other similar pollutants is not so serious insofar as analyses have not often revealed more than minute concentrations. In a few cases, particularly shellfish taken in some inshore areas, rather high concentrations of cadmium, zinc and lead are found. If eaten in conspicuously high amounts such fish could possibly cause risk to health in some individuals. On the basis of a review of cadmium and lead in foods, experts in the UK have concluded that the imposition of a statutory limit on the amount of the former is not justified; limits for lead already exist and these remain unchanged.

As well as causing toxicological risks, contamination with some elements can have damaging effects on flavour and appearance. Higher than normal concentrations of copper and zinc in shellfish like oysters result in an unpleasant metallic flavour and green colour. It should be noted that not all green-tinged oysters are contaminated with these metals. In some growing areas invasion of oysters by a greenish-coloured diatom (a kind of microscopic algae) gives rise to a desirable colour and savour.

Many countries are now taking voluntary or mandatory action to reduce pollution of the aquatic environment with heavy metals.

(ii) Organic chemicals

An enormous variety of chemicals from industrial processes find their way into the aquatic environment. Eventually minute quantities of some of them end up in fish. Of most importance are the persistent chemicals, that is those not broken down rapidly by natural processes. Notable among these are a group of chlorinated hydrocarbons including the insecticides DDT and its breakdown products [abbreviated as DDE and DDD (also known as TDE)], aldrin, dieldrin, benzene hexachloride (BHC or lindane) and poly-chlorinated biphenyls (PCBs – chemicals having a wide variety of industrial uses). Massive increases in the use of these chemicals over the past 40 years has led to increasing concentrations in all animal and plant tissues. The biological effects in nature have, however, been very slight though not unimportant. Decreases in bird, sea trout and salmon populations have been attributed to their use but no effect on the health of man has ever been demonstrated.

Nevertheless it has been deemed wise to reduce as far as possible the world burden of these chemicals and the use of several of the most widely dispersed chlorinated hydrocarbons is now banned outright or severely restricted in many countries. Where, as in the UK, these chemicals have been continuously monitored it has been possible to observe a reduction in the concentration in foods of, for example, DDT and dieldrin.

In the USA the Food and Drug Administration have imposed maximum permitted concentrations in most fish of 5.0 mg DDT and 0.30 mg dieldrin per kilogram wet weight. The joint Food and Agriculture Organization/World Health Organization Codex Alimentarius Commission have recommended similar maximum concentrations of pesticides in meat; fish is not yet included. However, it is unlikely that these concentrations would be found in the majority of fish and there is no reason why fish industries should be concerned to analyse routinely for substances of this kind. The only exception would be instances where quantities of fish were suspected of being contaminated with residues of chemicals. For example, in the very few areas of the USA having very high pesticide usage some fish have been found to contain more than the permitted maximum levels of these chemicals and in consequence have been seized by the authorities. Several countries have set up routine analytical monitoring for many chemicals of this group and these activities provide additional assurance about the safety of a wide variety of fish.

Very occasionally catches become grossly contaminated with chemicals, especially mineral oils, resulting from accidental or other kinds of large-scale release. Evidence of such contamination is apparent in tainted odour or flavour of the fish. Almost always tainting is detected during normal handling and processing and affected fish removed at this stage. No sensible merchant would risk jeopardising the health or goodwill of his customers by deliberately selling them tainted goods. As a further safeguard, authorities usually close down the fisheries in affected areas until the contamination has been cleaned up. Analyses indicate that the generality of fish contain no more than very low concentrations of hydrocarbons like those found in mineral oils. It is not certain that these hydrocarbons originate from oils released into the environment by man's actions. In any case there is no evidence that such hydrocarbons in fish present a hazard to health.

(iii) Radioactive isotopes

Exposure to high energy radiation originating from radioactive isotopes can be extremely injurious to health. Naturally-occurring radioactive isotopes occur in fish, as in foods, but at very low concentrations not considered to have a significant effect on health. In some areas artificial isotopes have been released in very small amounts into the environment, principally by Governments, and there is a possible danger of these accumulating in foods. The measurement and control of levels of radiation in food are responsibilities invariably assumed by Government under the coordination and advice of international agencies such as the International Atomic Energy Agency, the European Nuclear Energy Agency and the International Commission on Radiological Protection. The last named has published recommended standards for permissible intake of radioactive isotopes some of which may be present in fish.

These testing activities are conducted with extreme rigour and stringency and thus no additional involvement by the fish industries is necessary. The very small amount of radioactivity additional to the natural background arises from several isotopes released during testing of atomic weapons and in the effluents of atomic power stations and nuclear fuel processing plants. Wherever these releases occur samples of fish in the vicinity are tested routinely for activity. No cases of injury to health from eating fish contaminated with radioactivity have occurred.

(iv) Micro-organisms

Although most of the micro-organisms present on the outer surfaces, gills and in the viscera of fish caught in unpolluted waters are harmless to man, food poisoning can be caused by two micro-organisms that may occasionally occur. These are discussed in the section on microbiology (Chapter 5). Pollution of fish by raw or inadequately treated sewage can, however, introduce pathogenic organisms, i.e. capable of causing health injury to man.

Sewage contains two such types of faecal micro-organisms: bacteria and a single member of the virus family. The bacteria include the large group of *Salmonellae*, different members of which cause food poisoning, typhoid and paratyphoid, and *Shigella* which cause dysentery. Only a very few viruses are known to be incriminated; those responsible for the severely disabling disease infectious hepatitis and what is known as Norwalk gastroenteritis are the most important.

These micro-organisms are transferred via sea water and detritus to fish living within several kilometres of sewage out-falls or sewage-polluted rivers. No sewage micro-organisms are normally detectable in the open sea. As long as the micro-organisms contaminating the fish are either not allowed to multiply and spread the contamination or destroyed by adequate cooking or curing, no risk ensues. However, if the fish are eaten raw or semi-preserved outbreaks of illness are inevitable sooner or later. All the predisposing conditions for risk obtain specially for bivalve molluscs – oysters, mussels, cockles and clams – which are often harvested in estuaries or shores exposed to sewage pollution, concentrate micro-organisms because they are filter feeders and are frequently eaten raw. In fact most of the many cases throughout the world of illness traceable to fish polluted with sewage have occurred with these animals.

The special need to control the intrinsic microbiological quality of molluscs has been recognised for over 50 years and in many countries adequate preventative sanitary measures are now in existence. Apart from removing the prime cause of pollution two stages of control are used: first, identification of intrinsically safe molluscs and the reduction of contamination in them; second, checking the microbiological quality of raw material before distribution and sale.

The first stage may involve one or more of the following:

(a) testing overlying water and sediment in areas of harvesting for evidence of pollution. If an area is found to be polluted it may be closed to harvesting temporarily or indefinitely. If found to be clean, or at least microbiologically acceptable, it can be designated and publicised as such and unrestricted fishing allowed.

(b) removing molluscs from an area of proven or suspected unacceptability or from an untested area and re-laying them in a known clean area of shore or estuary. The efficacy of this treatment depends on the fact that molluscs will rid themselves of contaminating micro-organisms through constant self-irrigation within 1–5 days. It is, of course, necessary to ensure

Fig. 2.6 This cleansing plant has tanks for containing sea water and oysters, pipes and pumps for circulating the water, and header tanks illuminated with ultraviolet light for sterilising the water.

that a sufficiently large flow of clean sea water is available in order to avoid the animals re-infecting themselves.

(c) removing the molluscs as in situation (b) to tanks through which clean natural or artificial sea water flows. An efficient arrangement here is to recirculate the water through filters and to employ a means of sterilising it. The latter can involve treatment with chlorine, ultra-violet light or ozone. To expedite the process the circulating water is sometimes warmed. The mechanism of cleansing or depuration, as it is called is the same as in (b). There is evidence that depuration is not as effective in removing viruses as it is in removing bacteria.

The second end-product stage of control is usually considered essential in order to reduce risk of illness to minimal proportions. Checking is done by taking a representative sample of the batch of molluscs and carrying out on it a quantitative assessment of the numbers of faecal micro-organisms present. A large number of different types of faecal micro-organisms – and of pathogens – are potentially present in any given sample but it is far too expensive and time-consuming to determine more than one or two of them. It is therefore standard practice to select what is known as an indicator micro-organism which is invariably present in sewage and whose behaviour in the environment and in the mollusc is to all intents and purposes indistinguishable from that of the pathogens. The indicator organisms now being generally determined are either a group known as faecal coliforms or the overwhelmingly predominant member of the group, *Escherichia coli*. Determinations of either indicator give very similar results. The methodologies used are outlined in Chapter 6.

Public health authorities in different countries have set standards for the maximum numbers of indicator organisms that are permitted in molluscs considered acceptable for human consumption; these standards are in general agreement. Standards have been set also for the maximum numbers of indicator organisms occurring in a fixed volume of the water overlaying acceptable harvesting or re-laying areas.

Because of the history of great public concern that has surrounded the purity of shellfish, the control of microbiological standards, regulation and closure of harvesting areas and the oversight of testing are in all countries in the hands of official public health authorities, though actual fishing, harvesting, farming and purification operations may be conducted privately. In the UK, local bye-laws exist which govern the adequate cooking of commercial shellfish harvested in suspected waters.

8 Occasional peculiarities

From time to time individual or small numbers of fish showing abnormalities occur in catches. For the most part these are recognised and immediately discarded by fishermen but sometimes the fish are landed and enter normal trade. Observant port health or quality control inspectors at fish markets will normally condemn diseased, disfigured or grossly contaminated specimens about which there is any doubt. If suspicious or grossly abnormal fish or flesh inadvertently enter the factory or shop it is recommended that the inspector or operative also play safe and discard the whole sample.

Few diseased fish are encountered in normal trade because they succumb easily to predators in their natural environment. However, some fish are seen with externally-apparent diseased conditions like ulcerative dermal necrosis which affects salmon and

Fig. 2.7 This tumour on the cod is one of many different kinds of abnormality encountered.

salmonid fishes. Tumours, ulcers and nodules are also fairly widespread.

Physical damage to food fish caused by animal predators before capture is occasionally seen. Examples include attack marks of seals on salmon, of sharks on many species and of the small crustacean 'sandflea' on Pacific halibut.

Colour abnormalities occur which render fish unacceptable for some purposes. The flesh of cod and other gadoids is on rare occasions found to be coloured various shades of pink. This is due to the presence of the natural red pigments astaxanthin and zeaxanthin believed to be derived from unusual feed or some metabolic disturbance. Albino fish and a green discoloration observed in flounder (*Limanda ferruginea*) are further examples of the same kind.

Peculiarities of odour and flavour are fairly common. Perhaps the most well-known of these is an odour and flavour reported in cod and other gadoids, mackerel and chum salmon; it is variously described as 'blackberry', 'weedy', petrol, diesel, iodine, sulphide. Although of natural occurrence it is sometimes mistakenly attributed to contamination with mineral oil. The substance responsible has been identified as dimethylsulphide (DMS). This does not occur normally in most fish (except in exceedingly small amounts when it may contribute to the normal odour and flavour of some fish like clams) but may occur as a result of feeding on certain organisms. Species of planktonic bivalve molluscs known as pteropods have been implicated, notably *Spiratella* (or *Limacina*) *helicina* but others including *S. retroversa* are believed responsible. These pteropods contain dimethyl-β-propiothetin which is converted in the fish to DMS.

A low concentration of the odour may go unnoticed in iced fillets cut from affected fish but if the odour is sufficiently intense the flesh becomes unusable. In salmon the odour becomes pronounced when the fish is canned. The level at which customer complaints are likely to arise can only be judged from experience.

Some control over this particular difficulty can be exercised by the avoidance of areas and seasons known to give rise to high concentrations of the feed: Labrador, West Newfoundland and certain parts of the North Pacific in the spring and summer are known bad areas from this point of view.

Freshwater and brackish water fish including trout, channel catfish, milkfish, bream and tilapia species, and shrimps and prawns farmed inshore, occasionally suffer from an earthy or muddy odour and flavour which can reduce consumer acceptance. An iodoform-like odour in shrimps and prawns is fairly common and can be objectionable in high concentration. In neither case have the compound or compounds responsible been identified unequivocally though the earthy odour occurs particularly in fish caught in waters harbouring high concentrations of certain algae or microorganisms (*Actinomycetes*) that themselves have a similar odour probably arising from the presence of substances known as geosmin and methylisoborneol. Some reduction in the odour can be obtained by keeping the affected fish for a period in odour-free water but often the best method of control has to be by the removal of offending samples.

3 Quality deterioration and extrinsic quality defects in raw material

In the fish industry it is sometimes impossible to draw a clearcut distinction between raw material and products: what is raw material to a processor may be finished product to a retailer. Nevertheless it is useful to identify a distinct raw material as consisting of harvested fish that has undergone only the initial relatively simple stages of handling and processing. These stages would be (a) before port marketing (b) before further processing on board. For the purposes of this discussion therefore raw material is somewhat arbitrarily considered to be chilled or unchilled whole fish and shellfish, gutted or ungutted fish and unpeeled or unshucked raw shellfish; in every case raw uncooked material is understood. The condition of such raw material is basic to the quality of many products and may entirely determine it.

Quality deterioration and extrinsic quality defects in this material will be considered separately. In a third section the special problems associated with keeping fish and shellfish alive during distribution and retail sale will be considered. By quality deterioration is meant those natural processes of quality reduction that occur after harvesting and that are quite independent of man's deliberate intervention; extrinsic defects are quality reductions in the post-harvested material caused by man's deliberate or accidental actions. There is the possibility of a considerable degree of planned control over both quality deterioration and quality defects: all three sections include points that it is necessary to be alert to in exercising that control. In the main practical details of implementation are omitted and only principles described.

1 Deterioration

This will be considered under the headings (i) causes and effects (ii) factors affecting rate (iii) prevention or amelioration.

(i) Causes and effects

In raw fish deterioration takes two forms: microbiological and non-microbiological.

Micro-organisms are present on the external surfaces (including slime) and in the gut of fish but during life are kept from invading the sterile flesh by the animal's normal defences. The normal population, or flora, on fish consists of several groups, or genera, of micro-organisms. On death, the micro-organisms or the enzymes they secrete are free to invade or diffuse into the flesh where they react with the complex mixture of natural substances present. The numbers of micro-organisms in the flesh grow slowly initially but then increase rapidly. Their microbial enzymic action results in a well-defined sequence of changes in odoriferous and flavorous compounds. Initially, compounds having sour, grassy, fruity or acidic notes are formed; later bitterness and sulphide or rubberiness appear; finally, in the putrid state the character is ammoniacal and faecal. Not all the different genera of micro-organisms originally present on the fish are responsible for these changes. The exact sequence of changes differs between species and has not been fully elucidated. However, in the many marine species that contain the odourless compound TMAO (trimethylamine oxide), one prominent reaction is its reduction to TMA (trimethylamine) which possibly in conjunction with fatty substances is alleged to smell 'fishy' but certainly on its own is always recognised as being ammoniacal. As described later the gradual reduction in concentration of TMAO and increase in TMA have been used as chemical measures of spoilage. Elasmobranchs contain high concentrations of urea which is microbiologically degraded to ammonia. At later stages of spoilage micro-organisms through the agency of secreted proteolytic enzymes also attack the structural components, proteins, resulting in a gradual softening of the flesh.

A closely related contemporaneous sequence of changes occurs in the odour of the external surfaces and gills or organs (where these are present). These odours are more intense than those in the flesh and as described in a later chapter can be used as excellent indices of degree of spoilage.

What has been described so far represents the normal sequence of spoilage changes in most raw fish and shellfish. Occasionally a different sequence of changes occurs where storage conditions favour the multiplication of anaerobic bacteria (those that grow in the absence of air). Such conditions can exist, for example, where a mass of fish is packed closely together in a pound or in a tank of chilling medium. Wood tends to harbour this type of bacteria and fish placed in close contact with old wood structures often suffer from this kind of spoilage. It is characterised by the rapid development in localised parts of the individual fish of an obnoxious rotten egg-like odour; fish so affected are commonly called 'stinkers' or 'bilgy fish' and are always rejected for human consumption. 'Stinkers' of this kind should be distinguished from fish having the type of 'blackberry' odour referred to previously – there is sometimes confusion between the two.

In addition to changes in odour and flavour the continued action of micro-organisms affects the appearance and physical properties of several components of the body. The slime on skin and gills, initially watery and clear, becomes cloudy, clotted and discoloured. The skin loses its bright irridescent appearance, bloom and (in species with large and obvious scales) smooth feel and becomes dull, bleached and rough to the touch. The peritoneum becomes dull and can be progressively more easily detached from the internal body wall.

Micro-organisms are the most important agents of deterioration in raw, wet fish since they give rise to the particularly undesirable flavours associated with spoilage. Thus the control of deterioration is largely the control of micro-organisms. The micro-organisms inhabiting fish harvested in cold or temperate waters are what is called psychrophilic (tolerant to cold) in contrast to the mesophilic (tolerant to warmth) micro-organisms inhabiting warm-blooded land animals. Psychrophilic micro-organisms are less affected by chilling than mesophilic, which partly explains why chilled meat keeps longer under the same conditions than chilled fish. The other probable reason is that, after *rigor mortis* has set in, meat is in general more acid than fish; acid suppresses the activity of spoilage micro-organisms.

Non-microbial deteriorations are of two kinds: enzymatic and non-enzymatic.

The former arise in the first place from the large number of different enzymes naturally present in the flesh. In life these are engaged in normal processes such as tissue building and muscular contraction and relaxation but on death they become involved in predominantly degradative reactions. One of these reactions is the gradual hydrolysis during the first few hours of glycogen to lactic acid, resulting when the process is complete in a fall in pH from about 7.0 to 6.0–6.8 depending on species and the condition of the fish. The decline in pH is accompanied by the natural post-mortem stiffening called *rigor mortis*. It affects quality insofar as the texture of the flesh is rendered somewhat firmer and its tendency to lose moisture when pressed is enhanced. In either case the effects are usually only very marginal. In terms of microbiological spoilage it is an advantage as pointed out before to have the pH as low as possible for as long as possible. In practice, the generation by microbiological action of basic nitrogenous compounds like TMA and ammonia gradually raises the pH during the period after *rigor mortis* has passed off. In stale or putrid fish the pH attains about 7.5 or even up to 8.0 in some species.

The phenomenon of rigor just referred to is caused by another complex series of enzyme reactions lasting several hours or days depending upon temperature. Further reactions then render the flesh progressively more flaccid again. Rigor is also of some importance for raw fish quality in that fish flexed or squashed when in this condition suffer enhanced textural damage. Also, pre-rigor fish held at temperatures of 15–25°C (that is, below the cooking point) tend to undergo a strong contraction process (accelerated rigor) that can break up the flesh and so render it unacceptable in appearance or for further processing. Methods do not yet exist for achieving the complete abolition of rigor, but clearly circumstances that lead to quality damage from its effects should be avoided.

In meat such as beef or pork prepared from warm-blooded animals a phenomenon known as cold shortening can occur when pre-rigor flesh whilst still warm is chilled rapidly. Only a certain proportion of animals give rise to the condition. However, cold shortening can result in serious losses because the flesh may be too tough to eat and it may be unsuitable for processing. Most fish do not exhibit this phenomenon but it has been demonstrated in certain tropical species; the practical effects are, however, not serious.

Perhaps the most significant enzyme deteriorations are those that affect flavour. The compounds responsible for the desirable sweetish, meaty and characteristic fish flavours of different species are changed by the intrinsic flesh enzymes to more neutral-tasting compounds with the result that the fish become relatively more insipid. If this process of autolysis (self-digestion) then continues sufficiently far it is believed that in many species the concentration of one particular breakdown product, hypoxanthine, becomes high

Fig. 3.1 Four herring at different stages of spoilage; the specimen at the top is perfectly fresh, the others show progressive amounts of belly burst.

enough to contribute to the bitterness characteristic of unfresh fish.

The viscera (guts) also contain enzymes one main group of which is responsible during life for digesting food. On death these digestive, powerfully proteolytic enzymes attack the organs themselves and the surrounding tissues. The rate of attack is particularly great in fish that have been feeding heavily ('feedy' fish). In such fish the organs quite quickly become degraded to a soupy structureless mass and the belly walls are either digested away completely or so weakened that the slightest abrasion or pressure causes them to tear. This condition, known as 'belly-burst' or 'belly-burn' is seen most often in pelagic species. Visceral enzymes are also capable of penetrating to the flesh and causing additional quality damage there. The digestive enzymes of some shellfish like the crustacea shrimp, lobster and rock lobster are especially active and are able to attack the flesh of even moribund specimens. Also in dead crustacea, enzymes called tyrosinases can diffuse from the viscera into the adjacent tail meat and cause the formation there of black pigments that ruin appearance. For these reasons these animals should be kept alive in as full vigour as possible until just before processing if the best quality is to be obtained.

Of the non-enzymatic deteriorations the most prominent is the development of rancidity. In fish this is caused by the attack of oxygen on the chemically unsaturated fatty substances (lipids) contained in the flesh and other tissues. Fish in general have lipids of higher degree of unsaturation than most other foods and are therefore particularly prone to oxidative rancidity. The deterioration takes the form of the development of a linseed oil-like, painty odour and flavour, generally reckoned to be unpleasant by consumers. White fish always have a relatively low lipid content and although it is possible that some rancidity may develop in them in the raw state it is not detected because it is either too low in intensity or masked by other spoilage odours and flavours. In some pelagic species like herring, mackerel and freshwater trout, all of which have high lipid contents, rancidity has been detected during spoilage, though doubt has been expressed about the validity of this observation. The organ constituting the brown meat of crab is similarly affected.

During stowage all fish tissues slowly lose fluid the amount of which varies with conditions but which may amount to 5–10% of body weight after 10 or so days in melting ice. This fluid, or drip, carries away some of the flavorous compounds and so contributes to the general reduction in flavour. The effect is enhanced by leaching when the fish, as is normal, are constantly irrigated with ice-melt water. Loss on the one hand of weight through drip and on the other of flavour through leaching are both examples of quality loss but in many circumstances no effective means of controlling them are possible. Leaching can have the opposite effect of slightly improving quality by tending to reduce the concentration of undesirable flavours in spoiled fish.

In some cases it is impossible to attribute important quality deteriorations unequivocally to microbiological, enzymatic or non-enzymatic causes. Thus during spoilage the blood contained in the kidney, its associated vessels and main artery along the backbone gradually diffuses into the adjacent flesh. For some purposes like retail display of wet fillets the extent of the consequent discoloration may become unacceptable. The mechanism of blood release is unknown and although likely to be autolytic in nature other explanations cannot be ruled out. The flesh of absolutely fresh scampi (*Nephrops norvegicus*) is almost colourless. During spoilage pink coloured pigments originating from the hypodermis, the membrane between the shell and the flesh, gradually suffuses the latter. In addition, larger amounts of the coloured hypodermis are left attached to the flesh during shelling. Most commercial samples of shelled scampi meats are coloured slightly or definitely pink indicating slight or incipient spoilage. The mechanism of this colour change has not yet been elucidated. The appearance of the eyes of bony fish (not elasmobranchs) is a good guide to degree of spoilage. Very fresh fish have bright, convex (bulging) eyes with a dark pupil; as spoilage progresses the eyes become duller and greyer and pass from being flat to concave (sunken). Fresh fish flesh is translucent; stale flesh tends to be opaque. Again in both cases convincing explanations are lacking.

In practically all circumstances where deterioration is occurring in raw fish, microbiological, enzymatic and other effects will be proceeding concurrently and interdependently. Their relative importance will vary at any time but normally microbiological spoilage does not become significant until after the early period when intrinsic enzymes are active. Once microbiological spoilage is under way it increasingly dominates the picture.

With one exception spoilage *per se* in raw fish is not associated with dangers to man's health. The exception is a type of poisoning caused by eating frankly spoiled meat from certain dark-fleshed fish, a topic which will be dealt with in further detail in the section on microbiology (Chapter 5).

(ii) Factors affecting rate of deterioration

All the deteriorations under examination here are caused by biological, biochemical or chemical reactions and since all these (short of cooking) proceed faster at higher temperatures it is clear that raising the temperature by even a small amount increases the

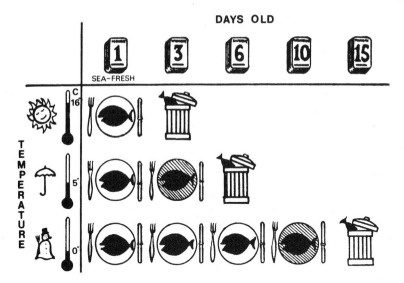

Fig. 3.2 The effect of temperature on the storage life of 'wet' fish is very marked; the storage lives illustrated are typical of demersal species caught in temperate or Arctic waters.

rate of deterioration. For example it has been found experimentally that increasing the temperature from 0°C to 5°C at least doubles the rate at which cod and similar species spoil. The large benefit obtainable from chilling and the fact that it is in many situations relatively easily accomplished, means that control of temperature is of prime importance in the control of quality deterioration in raw fish. The lower the temperature the better – as long as the fish does not freeze. One exception to this rule is abalone (*Haliotis* spp.) which after capture and during rigor rapidly loses fluid from the flesh and becomes tough. This process occurs minimally at about 6°C; at temperatures lower and higher than this the rate is more rapid. Incidentally, the toughness of this animal can be reduced by judicious hand pounding of the meat. Micro-organisms and enzymes can be rendered inactive by heating to a sufficiently high temperature but since this cannot be achieved without concomitant cooking, it is inapplicable as a method of controlling spoilage until cooked products are considered. In other words, it is impossible to pasteurise fish without effecting at least partial cooking.

Removal of the causative agents where this is possible has a large beneficial effect on the rate of deterioration. Clearly intrinsic enzymes cannot be removed without destroying the commodity. In order to be practically effective the numbers of micro-organisms have to be reduced to a small percentage of those originally present but this can be achieved in many cases by washing with cold water the surface of the fish free of slime, debris, gut fragments and faeces. In addition means can be adopted for preventing the contamination or recontamination of fish with extraneous micro-organisms, i.e. originating from other fish, structures, knives, ice, chilling media. The fact that the intestine and gut cavity of fish harbour large numbers of spoilage micro-organisms and autolytic enzymes renders it often of advantage to clean these away by the widespread practice known as gutting (evisceration). In some species this is always adopted because it has been found to reduce almost invariably the rate of spoilage and autolysis especially in those cases where the fish are stored for long periods. What is more, the removal of semi-liquified guts in order to further process the fish is an unpleasant, messy business and it is easier to carry out the process before storage. In other species gutting is avoided especially on board fishing vessels because either the fish are too difficult (e.g. spiny redfish – *Sebastes* spp.) or too small and numerous (e.g. sardines, herring) to handle in the time available. Where the feed is disgorged on catching or where fish are not feeding, gutting makes little difference to the rate of deterioration. Occasionally the gills are removed because they are alleged to be a source of microbiological and enzymic contamination but there is no evidence that this practice affects the rate of deterioration.

Enzymes and consequently micro-organisms can be inactivated or their activity suppressed by certain chemicals. In practice this device is limited by those chemicals that are permitted for use with fish and are recognised as non-injurious to health. Microbiological spoilage of the commodities under examination can be inhibited effectively by, for example, the addition of suitable concentrations of chlorine to washing and cleaning water, the storage of fish in an atmosphere of carbon dioxide or in sea water saturated with carbon dioxide, and the addition of antibiotics to the ice or the refrigeration medium used to preserve fish. The chemical control of autolysis is necessarily more difficult because the active substance must be able to penetrate to all or most parts of the fish. However, success in controlling the development of tail blackening in Crustacea is possible by immersing the animal in a solution containing sulphur dioxide or sulphites, both of which inhibit the tyrosinase enzyme responsible for black pigment formation. Success in this case is possible because the small size of the species allows rapid penetration of the chemical within a feasible time. Inhibition of lipid oxidation in fatty fish can be achieved in principle by preventing the attack of oxygen by means of treatment with chemical antioxidants, or storing in a gas tight, nitrogen-filled container or in a vacuum system. Some success has been achieved experimentally with the last of these.

A somewhat similar inactivation or suppression of enzymes and micro-organisms can be achieved by irradiating the fish with certain very high energy rays, in particular γ-rays. The effect on any particular sample of fish is related to the total amount of radiation: if the amount is low the number of micro-organisms is reduced somewhat, if high enough complete sterility results. The amount of radiation is measured as doses in units of 'grays'. This international unit has in the past few years replaced the previously used 'rad'. The relationship is that 1 gray equals 100 rads; normally, doses are expressed in kilogray (k Gy). On typical fresh fish a dose of 0.5–3 k Gy is sufficient to reduce numbers significantly; a dose of 5 k Gy and above will sterilise completely. The former is analogous to a partial sterilisation or pasteurisation and is accordingly called 'radurisation'. The storage life of initially fresh fish in ice can be extended two or three fold by radurisation and even more by sterilising doses – an apparently amazing prospect. Furthermore, the load of pathogenic micro-organisms can be partly or entirely killed. A dose of 2 k Gy, for example, kills most pathogens in fresh fish. Irradiation has therefore two potential benefits for fish: considerable extension of storage or shelf life, and reduction in risk to health. The same benefits exist for other foods. Unfortunately there are several major snags.

In the first place, irradiation also affects the substances in the fish itself and in many cases sterilisation by this method results in the development of off-flavours. Radurisation of white fish on the other hand can be accomplished without this happening: radurisation of fatty fish can result in oxidative rancidity and is therefore not recommended for these species. In the second place, the doses and kinds of radiation required are very harmful to man and special precautions and expensive, heavy equipment have to be used to eliminate totally the possibility of people being irradiated during processing. It should also be pointed out that the full benefits of irradiation are only realised if fresh fish is used and steps are taken to prevent recontamination of the fish with micro-organisms. These requirements would undoubtedly add fairly considerably to the cost of the overall operation and even if the process is economic may confine the equipment to a few large fishing vessels or a few large ports. In the third place, there is considerable popular suspicion, prejudice and political wariness in many countries about irradiated foods. Almost universally the marketing of irradiated foods is strictly regulated and so far only a few Governments have given permission for their use, and then only in restricted circumstances. There is a general view that it would be essential to label irradiated foods as such, and consequently to be able to test food to detect whether or not they had been irradiated. Currently, no tests are available that can unequivocally reveal whether the generality of foods have been irradiated. Until such tests are available it is difficult to see food irradiation being introduced on an unrestricted scale. Until 1981 there had been concern about the wholesomeness of irradiated foods, that is whether eating them could be injurious to health. However, in that year an Expert Committee drawn from the Food and Agriculture Organization, the International Atomic Energy Agency and the World Health Organization recommended that, based on exhaustive research, foods irradiated with doses up to 10 k Gy were safe to eat. Several Governments have since then endorsed this recommendation. Nevertheless taking all the impediments into account, the future for the irradiation of fish remains problematical.

It has been established that, other things being equal, large fish keep marginally better than small of the same species. This is not unexpected in view of the fact that the main mechanism of spoilage is the penetration of the causative agents from the surface to the interior of the fish; larger fish have a smaller surface to volume ratio so that in the same time less of their interior is affected. In addition, small fish of a given species tend to have a higher post-rigor pH than larger fish, which as pointed out previously is a condition conducive to greater microbiological activity. It is probable that much of the difference noted between the deterioration rates of different species is attributable to size differences, though differences in composition and microbiological flora may also play a part. Significant differences in spoilage rate exist between fish living in very different environments; the most notable example of this is the much longer keeping times under the same conditions that are characteristic of species caught in warm sub-tropical or tropical waters over species in temperate, sub-polar or polar waters. It has been tentatively suggested that the intrinsic microbiological flora of the latter have a more psychrophilic character than the former and so are less inhibited by chilling. It has been found recently that certain bony fishes and elasmobranchs living in deep marine waters (800–1100 m) keep for up to about 50% longer in melting ice than species caught in more usual depths; the reason for this behaviour is not known. Small differences in deterioration rates of similar fish of the same species at different seasons are well authenticated. Thus, N.E. Atlantic cod keeps best when caught in late winter and worst in mid-summer. These differences in seasonal behaviour are as yet unexplained but perhaps are caused by compositional or microbial flora changes related to the spawning cycle. Finally, small differences exist between the keeping quality of the same species caught by different catching methods. For example, cod caught by line and bait keep in general better than those caught by trawl: the latter method yields fish that are more damaged and more contaminated by mud and faeces. In all the cases covered in this paragraph knowledge of the phenomenon is of little or no help in designing practical means of control over the rate of deterioration. Either the effects are too small to bother about or are immutable parts of the fish industry and thus cannot be exploited constructively. Nevertheless, these differences contribute to the difficulty of approaching exact, determinative control over quality.

(iii) Practical means of preventing or ameliorating deterioration

As has been pointed out the most important factors in reducing the deterioration of raw material are time and temperature. Fish should be always chilled promptly to as low a temperature as possible and kept there for as short as possible. It is almost always possible to find ways of avoiding delays before the temperature of fish is reduced on board, when the fish is marketed or when it is transported. Those responsible for quality maintenance should explore these ways and make sure they are adhered to. The period between catching and marketing is, of course, determined by the type of fishing, distance of fishing grounds and speed of vessel; it cannot be changed unless these parameters are changed.

Quite simple, commonsense ways of avoiding rises of temperature or of reducing temperature suggest themselves. Fish should be protected from direct radiation from the sun, hot surfaces or heating appliances; they can be dowsed periodically with cool sea water; in fishing vessels the fish room should be insulated from heat gains through engine-room bulkheads; fish should not be allowed to warm up nor removed from the medium used to chill them until just before processing. It will be noticed that the practice in many market auctions of displaying in an unrefrigerated condition previously refrigerated fish violates the last point.

The simplest, most effective and often cheapest method of reducing the temperature is to surround the fish with crushed, melting ice. As a medium for absorbing heat, ice combines the virtues of being compact, clean, safe, flexible in use and relatively easy to apply. It should be in as intimate contact as possible with the fish and in sufficient amount both to cool the fish to very near 0°C and absorb incoming heat. Slowly melting ice through the washing action of the melt water keeps fish marginally longer than 'dry' ice and for this reason the ambient air should be slightly above 0°C. Stowage should be in well-drained, shallow layers of fish and ice; deep layers cause loss of weight and some damage to the fish. Wet strength paper is sometimes used in stowage to protect the whole fish from indentations and areas of bleaching produced in them by lumps of ice. This practice is not recommended because it often materially reduces the effective contact between ice, ice meltwater and fish. Better suffer the slight cosmetic defect than risk untoward spoilage. Alternatively ice of a smaller grain size or flakeice should be used. Given the choice small fish should be placed in ice before large ones. Further details of methods of applying ice and the recommended amounts to suit different circumstances are detailed in practical fish technology texts. Illustrations of how ice should be used in stowage, say, on board fishing vessels are given in Figures 3.3–3.7.

Properly used, ice can keep white-fleshed species of normal size caught in temperate or cold waters for up to 12–18 days before they become inedible to most tastes. Large fish, particularly those like halibut and tunas which tend to have a low post-rigor pH keep for 21–22 days. For up to 4 days the fish will be of top class or excellent freshness and between 4 and 15 days they will be in various states of decreasing freshness. For corresponding ungutted small pelagic species of low fat content and containing small amounts of feed these figures should be halved and for high fat content, 'feedy' fish, divided by three. Corresponding species caught in warmer waters,

Figs. 3.3–3.6 Several practical details should be observed when ice and fish are stowed on, say, a fishing vessel.

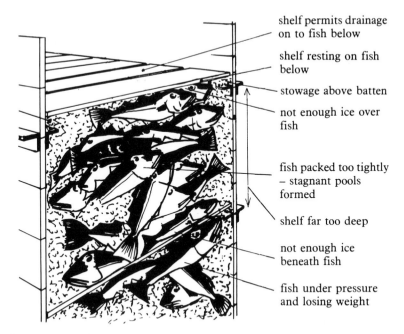

shelf permits drainage on to fish below

shelf resting on fish below

stowage above batten

not enough ice over fish

fish packed too tightly – stagnant pools formed

shelf far too deep

not enough ice beneath fish

fish under pressure and losing weight

Fig. 3.3 Incorrect bulk stowage.

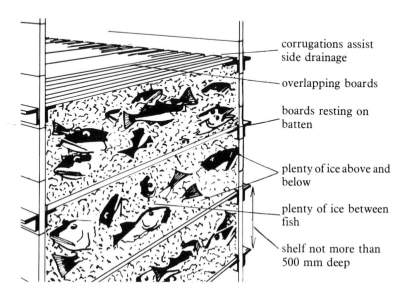

corrugations assist side drainage

overlapping boards

boards resting on batten

plenty of ice above and below

plenty of ice between fish

shelf not more than 500 mm deep

Fig. 3.4 Correct bulk stowage.

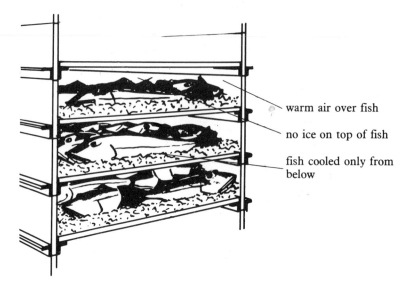

warm air over fish

no ice on top of fish

fish cooled only from below

Fig. 3.5 Incorrect shelf stowage.

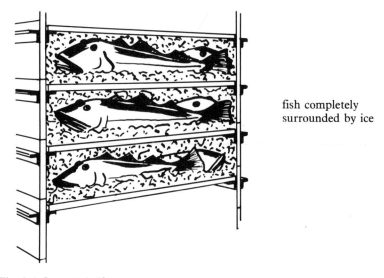

fish completely surrounded by ice

Fig. 3.6 Correct shelf stowage.

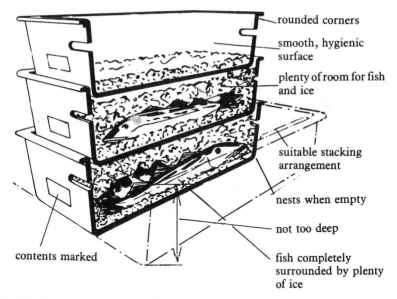

rounded corners

smooth, hygienic surface

plenty of room for fish and ice

suitable stacking arrangement

nests when empty

not too deep

contents marked

fish completely surrounded by plenty of ice

Fig. 3.7 Boxed stowage – good design.

both fresh and marine, will keep for periods in ice 50–100% greater than these. Most small, cold or temperate water shellfish like shrimp, prawn, scampi, abalone, scallop and clam will remain edible for up to 6–10 days; warm-water species for up to 8–12 days, whilst larger varieties will last somewhat longer. Other shellfish like oysters, cockles, mussels, crab, lobster and rock lobster are not normally iced in the raw state. An exception is a fishery, like that for lobster or rock lobster, in Brazil or Australia, prosecuted far from land-based processing facilities.

The storage lives of fish and shellfish when stored in melting ice can be summarised as shown in Table 3.1.

The advantages in terms of temperature control of retaining fish undisturbed in ice from the point of catching to that of further processing or even of retailing are obvious. This can be achieved by using containers of suitable size and design that are removable from and returnable to the fishing vessel. In some fisheries this practice is long standing, the containers being a wooden, light metal alloy or plastic box of 35–70 kg capacity. One major advantage of containers of this size is that they simplify considerably the problem of grading fish at the market or factory reception area for size, species and degree of freshness. Fish of the same size and species can be

Table 3.1 Storage lives of fish and shellfish when stored in melting ice.

Type of commodity	Number of days to the end of	
	High quality	Edibility
White-fleshed, average size, gutted or ungutted		
Caught in temperate and cold water	3–4	12–18
Caught in warm water	6–8	18–35
Large halibut, tuna and similar fish	5–6	21–22
Dark-fleshed, small fish, gutted or ungutted		
Low fat	2–3	6–9
High fat, containing large amount of feed	1–1½	4–6
Shellfish		
Caught in temperate and cold water	2–3	6–10
Caught in warm water	3–4	8–12

placed in the same boxes and assuming like boxes are grouped together into lots offered for sale on discharge of the vessel, the allocation of size grades to the lots can proceed straightforwardly. Similarly, the freshness of all fish in one box (or somewhat less surely, all the fish in lots of boxes) will be the same or very nearly the same and again allocation of a freshness grade is easy. Often discharging and marketing procedures tend to result in lots that contain fish mixed with respect to degrees of freshness: the allocation of grades to such lots is thereby rendered difficult or impossible. Large shipboard containers for stowage are also employed on a small scale but they present problems associated with handling, discharge and the avoidance of mixing. With care the average weight of fish in a box can be controlled manually accurately enough to satisfy merchants who buy the box. Robust balances suitable for weighing fish and ice into boxes on quite small fishing vessels have been introduced recently: these can provide useful assurance that boxes contain measured quantities of fish and that the ratio of fish to ice is appropriate. A common error when boxing fish in ice is to overfill. If, as so often happens, the boxes are stacked, the stack may be rendered unstable and the fish damaged by crushing. The overall economic advantages of boxing or containerisation at sea in terms of the quality advantage gained are not always self-evident and may require careful assessment.

An alternative method of lowering temperature is to immerse totally the fish in a chilled liquid medium. In practice this is either

Fig. 3.8 The fish laid out on the auction market remains mixed with ice in the box into which it was placed at sea; this practice has several advantages as far as the maintenance of quality is concerned.

plain water or more often sea water held in suitable tanks. Refrigeration can be mechanical (compressor and cooling pipes) or by adding ice to the medium. The use of ice is safer in that the mixture can never freeze but the choice will depend upon circumstances. Sea water can be chilled down to about $-2°C$ before it freezes and as long as the fish itself does not freeze, the extra lowering of temperature so provided can give a marginal preservation value. The necessary accurate control of temperature under normal conditions of varying heat load is, however, not easy to achieve in practice. It should be noted here that the process known as 'superchilling' is not chilling as normally understood: it will be dealt with in the section on frozen products. When using refrigerated sea water care should be taken through adequate mixing to avoid stratification in the tank and the development of 'hot spots', and the accumulation of adhering masses of fish which create conditions of anaerobic spoilage. Mixing can be achieved either by an adequate forced circulation system for the chilling medium or by forcing fine streams of air through the base of the container. Under proper control, refrigerated aqueous systems of this kind are capable of preserving the quality of all kinds of fish and shellfish approximately as effectively as equivalent storage in melting ice. However, there are two disadvantages that can affect quality: first, the colour of some species tends to become somewhat bleached on prolonged storage; second, the fish absorb salt from refrigerated sea water and if left in too long the flesh of small fish can become excessively salty. Salt penetration can be reduced by keeping the brine to fish ratio small. Among the advantages as far as maintenance of quality is concerned are the effective washing, descaling and (in the case of gutted fish) bleeding that are accomplished, lower weight losses, the reduced physical damage through crushing and a tendency to firm the flesh which aids further processing by machinery or canning. Ancillary advantages are ease of discharge by, for example, pumping or brailing, of handling large quantities of small species and of bulk transport to processing factories or markets. For this reason the method is particularly suited to the chilling of small pelagic species or of small shellfish. Insofar as these types of fish are almost always caught in large quantities over short periods, the problem of having a wide spectrum of freshness qualities within a large lot does not arise.

Chilled sea water has also been found to be of great practical advantage in improving and equalising the quality of fish frozen at sea. On larger freezer trawlers rather long delays can sometimes occur between catching and freezing. If left at ambient tempera-

Fig. 3.9 Herring being discharged from a shipboard tank containing refrigerated sea water.

tures fish subjected to these delays can suffer untoward spoilage, softening and break-up of the flesh, particularly during operations in warm climates. Chilling with ice to cover this eventuality is usually impracticable because of the amount of handling involved: refrigerated sea water provides an ideal solution. Another important advantage is that fish inadequately bled and then frozen yield pinkish or brownish discoloured flesh that is marked down in

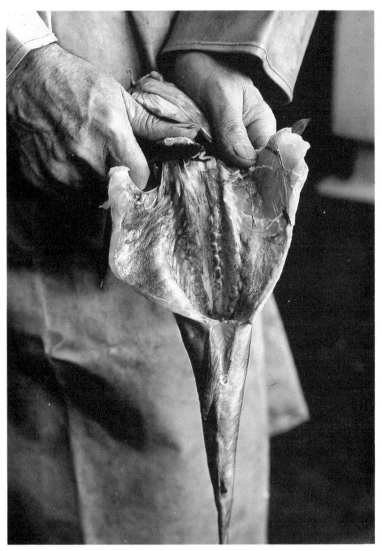

Fig. 3.10 An example of a well-gutted, double-naped and cleaned-out fish.

quality for some purposes. Thus, after gutting (which severs important blood vessels) a delay of 30–60 minutes preferably at chill temperatures is necessary to allow blood to flow out of the flesh. The required delay and rapid chilling are efficiently provided by pre-freezing immersion in refrigerated sea water.

Another application of chilled sea water is for highly-prized species such as snapper where retention of the best possible appearance and colour is an essential quality feature for the market. This is achieved by rapidly chilling individual specimens in a sea water ice slush and then packing in a plastic bag surrounded by ice. Incidentally, a further improvement in the retention of quality in these species is obtained from the practice of killing by the rapid insertion of a metal spike into the brain.

Chilled sea water can also be sprayed over fish laid out on shelves, but the technique is very much less widespread than those just described.

Practical measures to improve cleanliness and reduce contamination are for the most part self-evident. Guts should be removed completely and cleanly soon after capture; the practice of double napeing round fish should be encouraged (that is, cutting through both sides of the belly wall anterior to the pectoral girdle). If possible, heads (cephalothorax) should be removed similarly from

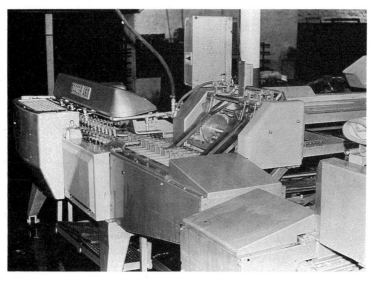

Fig. 3.11 Machines such as this modern one for nobbing, are available for rapidly and efficiently removing guts and heads.

shrimps and prawns. Machines used for gutting should be adjusted properly and their cutting devices kept sharpened. Guts and their contents should not be allowed to contaminate as yet ungutted fish. Washing of fish by hand or machine should be thorough using clean fresh or sea water. Dirty water should be drained off the fish after washing. Knives, pounds, containers, the deck and its fittings, and the bilge should be kept clean and if possible a regular routine of cleaning and sanitation of the vessel instituted between voyages. Similar considerations apply to the market, its environs and equipment, and to vehicles. Surfaces should be robust, impervious and easily cleanable. Because it is very difficult to clean and can cause the occurrence of 'stinkers', wood where it is likely to come into contact with fish, should be avoided. Similarly, open-mesh containers or structures made of natural fibres such as sisal, straw or wicker are undesirable. Smooth metal alloy or plastic are the materials of choice. Less apparent is the need to use clean, fresh ice every time fish is stowed. Fresh ice, assuming it is made properly from good quality water, has a low bacterial load, but on storage especially in fishing vessels or in contact with fish, the numbers can greatly increase. Old ice should be discarded after use or after a voyage if it is not used up. Furthermore, chilled sea water should not be allowed to become too contaminated with blood, debris and faeces; after use it should be thrown away.

Although it is true that the chemical substances mentioned previously can exert significant preservative effects, in practice the benefits in terms of longer keeping times are outweighed currently by their additional cost and inconvenience. Chemical preservatives do not help to retain absolutely fresh character but delay the onset of gross microbiological spoilage. More benefit can often be achieved by freezing. Furthermore, there is a growing mistrust of the use in foods of chemical additives or preservatives of all kinds and a continuous pressure to reduce their number. As examples of effective preservatives that have lost favour one can quote the antibiotics oxytetracycline and chlorotetracycline. By the addition of low concentrations (about 5 ppm) of these substances to ice or refrigerated sea water used to store cod, an extension of keeping time of about 4 days can be obtained. For a short time these substances were used in small quantities in some fisheries but do not appear to be used now anywhere. By storing chilled whole fish in an atmosphere of carbon dioxide a similar extension of storage life can be obtained. This has been achieved in large-scale trials by either

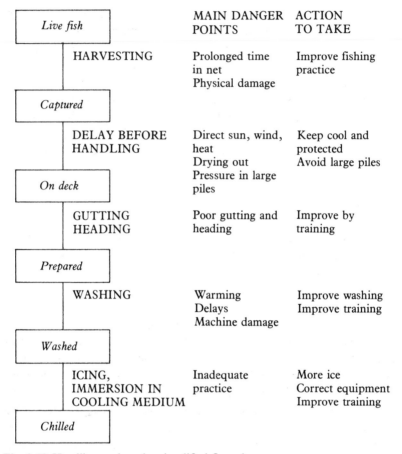

Fig. 3.12 Handling on board – simplified flow chart.

placing fish and ice in a gas-tight container into which carbon dioxide can be fed, or by bubbling carbon dioxide through fish held in refrigerated sea water. The technique is, however, rarely used for the preservation of bulk raw material because of difficulties of control and handling; also some species tend to become bleached and softened after long storage under these conditions. In summary, it can be said that as far as is known no substances capable of controlling microbiological spoilage in bulk raw material are used commercially at the present time. The use of sulphite is permitted or accepted by common usage in some countries as a means of controlling tail blackening in Crustacea but this is control of an enzyme rather than of micro-organisms. From time to time proposals for new chemical preservatives are made but none have come to fruition or are likely to. Even if any were phenomenally successful the cost of carrying out the elaborate tests necessary to prove their safety in use and the strong current of opinion against the use of preservatives in food would weigh heavily against them.

A considerable extension of storage life can be obtained if chilled fish are held in an evacuated pressure vessel. This so-called hyperbaric storage, like bulk carbon dioxide storage, has not been taken up commercially because of comparable disadvantages.

2 Defects

Varying degrees of quality loss in raw material can be caused by human carelessness or ignorance and almost all are preventable. Several examples of defects and their correction have already been pointed out – bending during rigor, poor gutting, napeing, cleaning or washing.

The importance of bleeding has been emphasised in connection with treatment before freezing, but this also applies in many instances to ordinary chilled fish. Customers buying chilled white fish in many parts of N.W. Europe, for example, customarily expect the flesh to be very pale. The occurrence of blood clots, dark patches or an overall darkening of the flesh is considered defective, and depending on their incidence these defects reduce the value and acceptability to varying degrees. Similar defects must also be absent in salmon. In Norway regulations lay down bleeding regimes for fish. The occurrence of discolorations of this kind can be reduced by thorough bleeding after capture. Fish blood in vessels and organs remains fluid for up to about 30 minutes at chill temperatures, but tends to clot rapidly after this time or sooner at higher temperatures. Therefore in order to achieve the best result, fish should be cooled immediately after capture, bled within about 30 minutes and allowed to bleed freely thereafter. In practice, ideal conditions of this kind may not be attained but the rate of fishing and handling the catch should be regulated as far as possible to allow them to be approached. In many cases adequate bleeding is accomplished through good gutting practice but it can be aided by cutting the throat or the tail off especially when the fish are alive or have just died. Marked improvements in the colour of the darkish-fleshed species saithe (*Pollachius virens*) are possible by effective bleeding. It is standard practice with farmed salmon to bleed either while the fish are still alive or after they have been asphyxiated with carbon dioxide.

As far as possible fish should not be crushed by equipment, ice, human feet or themselves. Piles of fish a metre or more in depth are still common sights in the fisheries of the world; such conditions are undesirable. Violent throwing about or dropping from a considerable height should be avoided. Care should be taken to prevent damage arising during pumping, fluming, or mechanical conveyance. Large catches in nets should not be hauled inboard in such a way that the fish is crushed; small lots should be brailed or pumped out of the net whilst it is held in the sea against the side of the fishing vessel. Gaffing or pitchforking as a means of moving fish is fairly common in some countries: it is an undesirable practice because many fish are inadvertently pierced through the edible part rather than the head. Efforts should be made to discourage it. Any bruises, abrasions or cuts spoil appearance, reduce yield of edible flesh and accelerate spoilage.

Fish should not be laid unprotected on floors of markets, factories, etc. or on beaches. The practice is unhygienic and can cause physical damage to the fish. Similarly, fish should be protected against being soiled by birds, insects, rodents, etc.

Incorrect or slovenly sorting on board or at discharge of fish by species, size and freshness can result in lots that are defective in quality insofar as they may not meet specified requirements of homogeneity. To take a simple example, buyers of cod do not want an admixture of haddock in their purchase.

It is fairly common for unwanted odours and colours to transfer from one species to another. The ammonia generated in the spoilage of elasmobranchs may contaminate bony fish if they are stowed with them. The pigments in the coloured spots on the upper surface of plaice can be transferred to the white underside of adjacent fish. In both cases the correction of the defects is to stow the fish well separated from one another.

Preservatives for wooden structures in fishing vessels should be selected with care because the odour from some can contaminate stowed fish; recommended non-tainting preservatives are available from reputable suppliers.

Fish exposed to sun or wind quickly lose their bloom and their appearance may become irreversibly damaged through excessive drying of the surface. Adequate protection is the obvious remedy.

Excessive salt penetration of fish held in refrigerated sea water can only be avoided through experience. As a guide, fish the size of herring or mackerel become unacceptably salty after 5–6 days in sea water at 0°C. The period of storage of salt content of the medium should be reduced.

The presence of unwanted or illegal substances or preservatives in raw material is obviously a quality defect that should not be allowed but unfortunately it is not always easily detectable. Clearly, routine or even occasional analyses for all or some of the many substances that might be used is almost out of the question and in fact unnecessary in view of the small risks involved. The best safeguard is to deal with reputable suppliers or to employ reputable workpeople. Otherwise occasional unannounced inspections of suspected procedures (where that is possible) and the operation of sanctions against the offender offer the degree of control required.

3 Keeping fish and shellfish alive

A part of the commercial trade in a very small number of true fish species is in live animals. Although reasonably common a century ago the practice of selling live white marine fish, especially cod, has now almost died out. On the other hand a relatively large proportion of the trade in the freshwater species, eels and carp is in the live animals. A small part of the trade in trout also deals in the live animals. To ensure survival of a reasonable proportion of the stock for periods of several months up to the point of sale, they require keeping at proper densities in cool, well-oxygenated and filtered fresh water. For short-term storage of a day or so during transport or display, carp can be kept alive wrapped in loose, moist packing material; for the same purpose, eels are held in special trayed containers designed to keep the animals moist and cool in a shallow pool of water. Further details are available in specialist literature.

It is regular practice to keep alive until sale larger varieties of Crustacea that is, crab, crawfish (spiny or rock lobster), lobster and squat lobster. The reason is that very rapid spoilage of these creatures sets in after death and keeping them alive is the only guaranteed way of presenting unfrozen, raw material of high quality to the customer. Preferably, only lively undamaged animals should be used as raw material for high grade products: moribund or damaged specimens should be killed and processed immediately.

Near the points of catching, these varieties are held alive either in containers moored in the sea or shore ponds constantly renewed with fresh, clean sea water. For prolonged storage of lobsters and lobster-like species tanks are used containing natural or artificial sea water that is filtered and oxygenated; crabs do not keep well in storage tanks and are not often held in this way. When held in groups, large Crustacea tend to fight amongst themselves and may damage or even kill one another. To prevent this happening crabs should be kept as closely packed together as possible and the claws of lobsters and their relatives immobilised by fitting bands round them or, less preferably, by inserting special pegs. These varieties can be kept alive out of water during transport for a day or so. For this purpose the wooden boxes and barrels previously used are being replaced by ones made of ventilated fibreboard or plastic, the animals being packed in wet seaweed, sawdust, wood wool, straw or paper to prevent damage and absorb excrement. Very good reduction in mortality during transport of lobsters is achieved by the use of sea water at about 5°C.

Hardy shellfish species, some of which are subjected to the depuration process described earlier, will be alive when despatched to customers from the cleaning plant or port. Included here are the molluscs: abalone, clams, cockles, mussels, oysters, scallops, whelk and winkle. Many indeed will be still alive when cooked or further processed, but being in this condition is not a necessary prerequisite for best quality because good products can be made from very fresh but dead animals. Dead bivalve molluscs can be recognised by the fact that their shells separate and it is a customary safeguard against poor quality to discard any that exhibit this condition.

4 Quality deterioration and defects in products

By 'products' is meant simply all those commodities, with the exceptions noted in the Preface, not covered in the preceding section. These have been arranged into eight groups each containing products of similar type. Processing, distribution and marketing will be all encompassed. In each case active control of quality is possible in principle.

1 Chilled processed fish for direct selling

Some unfrozen fish from port markets or points of landing is taken to merchants or processing factories and thence distributed to retailers, caterers or inland markets using only chilling as a means of preservation. The processing applied is often only gutting, filleting, peeling or shucking. Some whole gutted or ungutted fish is distributed as such to retailers who may carry out their own cutting or filleting operations. Also large quantities of fish frozen at sea either in the whole form or as fillets are thawed out on shore usually after intermediate frozen storage and thereafter is treated in exactly the same way as fish landed in the chilled form. In many developing countries the systems of preparation and distribution are often much simpler. For example, fish may not be gutted or treated at all between the points of harvesting and of sale. On the other hand, in developed countries there is an increasing development towards incorporating chilled fish and shellfish into sophisticated or elaborate products. These products include fillets coated with breadcrumbs and similar materials, fillets or shelled Crustacea in a wide variety of sauces, and fish or shellfish in pastry or salads. They all aim to provide the purchaser with a complete meal or main part of a meal requiring no preparation by the purchaser beyond cooking. The control of quality of such products is quite elaborate and clearly goes beyond the fish component itself. In all the above cases the factors to be taken into account in controlling quality of the chilled fish are basically the same as those dealt with in the previous chapter on raw material. In particular the pattern of spoilage is the same or very similar. It should be noted that the storage life of fish prepared from previously frozen and then thawed raw material is, to all intents and purposes, the same as that prepared from equivalently fresh unfrozen material, always assuming, of course, that the freezing, frozen storage and thawing have been carried out properly. Fish in merchants' or factory premises will almost inevitably undergo a good deal of further handling and exposure to temperatures higher than ideal, and thus the potential for deterioration will be greater. Additionally, fish cut into slices, steaks or fillets will be more susceptible to spoilage because it has a greater surface area; microbiological and other contamination will be greater and the product will tend to warm up more quickly.

It is, of course, obvious that many faults in the quality of raw material cannot be corrected at the further processing stage. Loss of freshness, for example, cannot be restored and as far as this is concerned, product quality is only as good as raw material quality.

Normally, chilled fish will be packed and distributed in easily handled boxes or containers of up to about 50 kg capacity made of wood, fibreboard or plastic, of light construction and often non-returnable. For display purposes, for example in shops, fish will often be removed from these containers and may be kept chilled by placing on a bed of ice or in a specially designed mechanically refrigerated chill cabinet.

(i) Controlling deterioration

Fish subjected to unavoidable delays before or during processing can be reduced to or maintained at a low chill temperature most effectively by mixing it intimately with crushed ice. The use of bulk storage in chilled liquid media during either processing or distribution of products is often impracticable. It is normally impossible to reduce sufficiently rapidly the temperature of a solid mass of fish (especially fillets) by placing it in a refrigerated chill room. Such rooms should be used in conjunction with ice for the stowage of chilled wet fish; their temperature should be at 2–3°C to allow a slow melting of the ice. Where ice cannot be applied, fish can sometimes be kept cool by immersing them in running water or being sprayed with cold water; however, prolonged immersion of fillets in water with resulting waterlogging should be avoided. In addition to these measures, the product should not be subjected to high ambient temperatures or direct radiant heat.

Most fish will be left unprocessed until after *rigor mortis* has passed off. But in some inshore fisheries and during freezing at sea, fish may be filleted whilst they are in the pre-rigor state. Under these conditions the process of rigor continues in the fillet. If the fish from which the fillets were cut is in good biological condition, rigor is manifested as an overall shrinkage, a roughening of the cut surface and a loss of fluid. In extreme cases and in practice fortunately rather rare ones, the length of the fillet may shrink by 30–40%, lose 25% of its weight and the surface become corrugated like crepe rubber. This problem can be overcome by keeping the fish a day or so in the whole state until they are in or through rigor. Alternatively, if the fillets are to be frozen, the fish should be kept chilled throughout and handled expeditiously in order to prevent the fillets going into rigor.

The ratio of ice to fish packed in containers for distribution should be adequate to reduce the temperature of the fish quickly and to maintain it near 0°C throughout its journey. This ratio has to be adjusted to take into account the prevailing ambient temperature and type of container (for example, whether insulated or not). Again, ice should be intimately mixed in layers with the fish and disposed in such a way that heat entering the container is absorbed before it reaches the fish. Where containers holding fish and ice are stacked for distribution in lorries, railcars and the like, care should be taken that the fish in the outer layers of containers is kept at 0°C throughout the journey under all climatic conditions: checks on fish temperatures with an appropriately designed and used thermometer or thermocouple particularly at the termination of journeys should be carried out periodically. Similar temperature checks are recommended at any vulnerable points in the harvesting, distribution and marketing stages. The use of expanded polystyrene containers for transporting fish and fish products is widespread; because of the useful insulation properties provided, their use is recommended. Wet strength paper is sometimes used to protect the fish from indentations caused by ice lumps and possible contamination. This practice is unexceptionable as long as it does not interfere with the job the ice has to do; with clean boxes and clean flake-ice it is unnecessary.

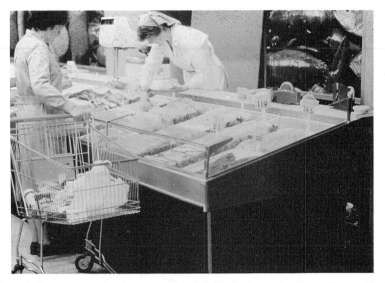

Fig. 4.1 Attractive packaging and good chilled display maintain quality and help to improve sales.

As far as retail display is concerned, it is not always easy to reconcile the conflicting requirements of keeping the fish properly chilled and moist, and of attractive arrangement. The skilful use of crushed or flaked ice coupled with some kind of transparent protective cover that will shield the fish from dirt and draughts remains the best choice. Mechanically refrigerated chill cabinets or display slabs are a reasonable alternative though they tend to dry fish because the ambient humidity is low; also control of temperature in them is not always easy and damaging partial freezing of the product can occur. Fish should not be displayed in deep piles: a maximum depth of about 5 cm should be aimed at.

Delays during which fish are held awaiting further handling or distribution can build up especially in large, complex factories and distribution chains but every effort should be made to keep them to a minimum. It is good practice to institute regular reviews of processing procedures to ensure that delays in production can be identified and the situation rectified. Stocks of fish in shops and catering establishments should be kept to a minimum and not regularly held over.

Cleanliness, hygiene and sanitation of premises and equipment should receive special attention. Guts, filleting and trimming offal, shell and other debris should not be allowed to contaminate the finished product and should be rapidly removed from its vicinity. It is often advisable to chlorinate the water supply in factories or processing areas of any size. Further special points will be dealt with in the later section on microbiology.

As with raw material it is possible under given conditions to extend the storage life of fillets, and of peeled or shucked shellfish by chemical treatment, by irradiation or by packing in gas-tight plastic pouches either evacuated or containing carbon dioxide. The chemical treatments include dipping in or spraying with solutions of antibiotics or so-called chelating agents (for example, ethylene-diaminetetraacetic acid). In practice the disadvantages of these adjuncts or alternatives to chilling currently outweigh the advantages in the main, and few are being used. Probably the most important current chemical treatment is the use of gaseous carbon dioxide in what is known as modified or controlled atmosphere packing. In this technique very fresh fish or shellfish, which may be gutted, filleted, steaked or shelled, are placed in a bag or tray made of thin special plastic suitable for retail display. The air in the bag or tray is displaced by a mixture usually consisting of nitrogen, oxygen and carbon dioxide, and the bag or tray then hermetically sealed. Extension of shelf-life of about 50% can be achieved, but

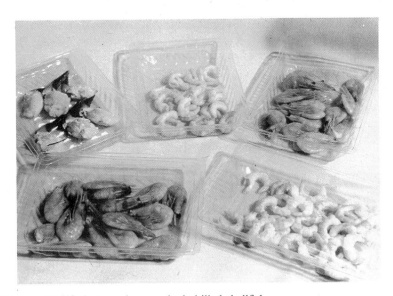

Fig. 4.2 Modified atmosphere packed chilled shellfish.

only if the storage temperature is maintained at between 0°C and about 3°C. At higher temperatures of storage the benefit is lost. The effect is due to the bacteriostatic action of the carbon dioxide but the pure gas cannot be used because appearance then suffers and drip can be excessive. Different gas compositions have to be used for different fish species to secure the maximum benefit. The whole process is done mechanically and requires sophisticated control. As mentioned earlier there is some minor use of sulphites to prevent blackening in chilled Crustacea.

During distribution, storage and display chilled fish especially in the form of fillets, loses fluid, or drip. The phenomenon is most marked with fillets or steaks cut from fish previously frozen and thawed, even when these processes are carried out properly. Depending upon the circumstances up to about 5% of the initial weight may be lost. Complaints from customers about underweight on delivery caused by this kind of loss can be taken care of by adding sufficient initial overweight but this can be expensive. The cut surfaces of fillets prepared from frozen and thawed fish tend to have a poorer bloom and appearance than corresponding fillets cut from never-frozen fish. Moreover, the unsightly fluid can accumulate in packages and so reduce customer appeal. All these quality losses can be corrected or disguised by dipping briefly (1–2 minutes) the fillets immediately after cutting in a 5–10% aqueous solution of polyphosphate – sodium or potassium pyrophosphates, sodium or potassium tripolyphosphates or sodium hexametaphosphate have all been found to be effective for this purpose. These substances act by swelling the outer surfaces of the fillet so partially sealing them. An additional benefit from this treatment is the glossy translucency imparted to the cut surfaces. The flavour of the cooked fish is virtually unimpaired. A similar but poorer effect can be achieved by dipping in concentrated brine but this is not recommended because the fish often become too salty.

Fish and shellfish are treated on a small scale with flavouring agents, the aim being to restore something like the fresh flavour that has been lost as a result of handling, washing, poor freezing, etc. Although mostly applied to frozen products, chilled products have also been treated in this way. Citric and ascorbic acids and monosodium glutamate are added to small, peeled Crustacea, and hydrolysed vegetable protein mixtures to white fish. Whether these treatments are really justified in terms of improvement in quality is debatable.

(ii) Defects

These can arise in either the preparation for packaging or the packaging itself. Taking preparation first, where cutting operations are included it is necessary to see that these are carried out correctly and that faults in workmanship do not occur. Points requiring attention in this respect are neatness and smoothness of filleting, complete removal or trimming (where these are called for) of skin, scales, belly membrane, viscera, bones, shell, fins, bruises, blood clots and flesh discolorations. Parasites and other abnormalities have been discussed earlier. Defects of this kind are detected visually during the cutting operations, at an inspection point or on a conveyor. The presence of harmful bones in fillets is detected by feeling with fingers at those places where the bones are known to protrude.

In addition to these points, adequate cleaning, washing and draining of the product should be ensured. In achieving the required standards, the importance of proper instructions to staff and adjustments of machines is evident. Care should be taken to ensure that pre-packing treatments are carried out correctly. For example, overtreatment with polyphosphate results in a slimy, unpleasant article. Of course any material possessing unusual appearance or off-odours should be discarded. Solutions used in treatments should be kept cool. Battering, coating with breadcrumbs and flash-frying to set the coating should be controlled so that the result is a uniform, even and fully adherent coating. Portions covered with specks of burnt crumb or containing excessive quantities of overheated fat should be discarded. The oil or fat in frying should be checked periodically for wholesomeness and absence of excessive discoloration or acid content. Fat stains in the wrapping material should be eliminated by adequate drainage and cooling of the product before packaging. The proportion of coating to fish should be kept within the specified limits. For elaborate products, checking for defects in the non-fish components such as sauce, vegetables or pastry needs to be done. In products like fish cakes, burgers or balls, the proportions of the ingredients being used should be checked from time to time by supervisory staff. Apart from the need to adhere to good commercial practice, in some countries the amount of fish in some products is controlled by legislation.

Turning to packing, overfilling or underfilling of containers should be avoided since both may lead to damage to the product

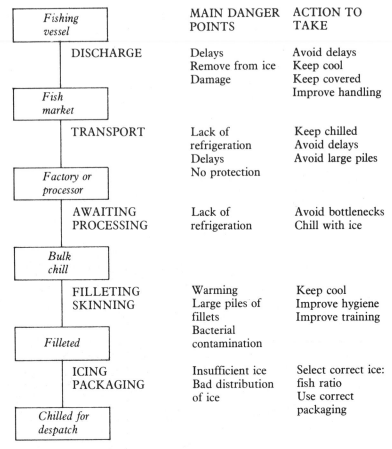

	MAIN DANGER POINTS	ACTION TO TAKE
Fishing vessel		
DISCHARGE	Delays Remove from ice Damage	Avoid delays Keep cool Keep covered Improve handling
Fish market		
TRANSPORT	Lack of refrigeration Delays No protection	Keep chilled Avoid delays Avoid large piles
Factory or processor		
AWAITING PROCESSING	Lack of refrigeration	Avoid bottlenecks Chill with ice
Bulk chill		
FILLETING SKINNING	Warming Large piles of fillets Bacterial contamination	Keep cool Improve hygiene Improve training
Filleted		
ICING PACKAGING	Insufficient ice Bad distribution of ice	Select correct ice: fish ratio Use correct packaging
Chilled for despatch		

Fig. 4.3 Filleting and chilling – simplified flow chart.

during transit. Checks should be made to see that the correct weight, including overweight where this is necessary, and the correct species are packed. Sometimes customers require uniform sizes or portion weights; in these cases care should be taken to see that grading and packing have been carried out according to the requirements. Visual appeal is a most important aspect of quality and the contents of packages should be arranged attractively, particularly in the case of retail pre-packages. Skin pigments can be transferred to cut surfaces; packing should avoid this.

2 Frozen fish

Freezing is a means of arresting either partially or completely the deteriorative actions of micro-organisms and enzymes. Essentially it is a type of partial, gentle dehydration in which the water is removed as ice. Micro-organisms cease to multiply below about $-10°C$ and the activity of enzymes (either microbiological in origin or intrinsic to the fish) is in general rapidly reduced or re-directed as the temperature is reduced below the freezing point of about $-1°C$. This means that the deteriorations suffered by frozen fish are qualitatively different to those discussed so far. Although some

Fig. 4.4 Freezing fish at sea; the growth of this practice over the past three decades has been a major factor in the greater control of fish spoilage.

are undoubtedly caused by enzymes, the nature of others is still obscure. Most deteriorative processes occurring in frozen fish are the same regardless of the product and therefore will be discussed in general terms. Defects on the other hand tend to be specific and will be dealt with more in relation to particular products. Many of the general remarks made earlier on damage, avoidance of contamination and the like apply, of course, equally to frozen fish.

Freezing cannot reverse deteriorations that have already occurred. Fish with a certain degree of pre-freezing spoilage will retain it throughout freezing, frozen storage and thawing. To that extent, controlling the quality of frozen fish includes controlling or selecting the quality of the raw material and controlling the preparation of fish for the freezing process. Thus, all the considerations previously described that affect the quality of both raw material and processed chilled fish have to be taken into account with respect to frozen fish. Frozen whole, round white fish, including that produced at sea, should be frozen within 2–3 days of death (storage being near 0°C) if good quality thawed products suitable for subsequent filleting and processing are to be obtained. For most flat fish this period can be 5–6 days and for small pelagic fish should be only 1–2 days respectively. The requirement for fish frozen as single fillets or as blocks of fillets to be used without thawing, is somewhat less demanding. For example, 5–7 days for round, white fish can be allowed. To obtain good quality, most shellfish in shell or not, after catching should not be kept in ice for more than 2–3 days before freezing. Not more than a day is recommended for abalone.

As with chilled processed fish, frozen fish is incorporated into a great variety of frozen products, including whole meals. The control of quality of such products again involves much more than the fish components and only some salient points can be made here.

(i) Deterioration

During the frozen storage of fish off-odours and off-flavours gradually develop. At first they are so slight as to pass unnoticed by most people, but after a sufficiently long period of storage they become so intense that the fish are almost always rejected as too unpleasant to eat. The exact nature of these deteriorations depends on the species but most particularly on whether or not it is a fatty species. The very characteristic off-odours and flavours assumed by lean fish and shellfish are variously described as acid, bitter, turnipy, cardboardy, musty or singed; those of fatty fish are not unexpectedly typically rancid, oxidised, painty or linseed oil-like. The latter are quite evidently the result of the oxidation of the lipids; those of lean fish may be. The fat on the surface of whole, thawed fatty fish when in an advanced state of rancidity feels gummy to the touch.

In parallel with these changes, the texture as sensed by hand or, in the case of the cooked product, by mouth gradually changes from the usual soft, springy, moist succulence of fresh or recently frozen fish to unacceptably firm, hard, fibrous, woody, spongy or dry. Whereas little fluid can be expressed from raw fish, copious amounts exude or can be pressed out after frozen storage under poor conditions. A further manifestation of the same change is that when fish that has deteriorated in cold storage is smoked the attractive glossy pellicle formed during normal curing becomes increasingly difficult to obtain and eventually only a totally dull, matt article results. The necessary tendency to become sticky when ground with salt and to form a gel satisfactorily on steaming as in the manufacture of fish balls, fish sausage or kamaboko is also gradually lost and the fish becomes completely useless for this purpose. All these textural changes are caused by an essentially irreversible phenomenon known as denaturation suffered by the flesh proteins.

Appearance also suffers. White fish and shellfish become opaque and yellowish; fatty fish develop a 'rusty' appearance; the pigments of fish and shellfish tend to fade and become duller or change in hue; 'bloom' gradually disappears.

The effects are enhanced if the fish are allowed to dry out – which tends to happen naturally during frozen storage. Dehydration is bad in itself because product weight is lost but equally if not more serious, the surface and thin parts of the fish become irrecoverably dry and porous, typically described as like balsa wood. This condition, somewhat misleadingly known as 'freezer burn', renders products quite inedible.

The rate at which these changes occur depends very much on temperature. At $-30°C$ where over 90% of the water in the fish has been converted into ice, they happen very slowly; fresh white fish will keep in good condition at this temperature for up to 8–9 months and will only become inedible through bad texture and off-

Fig. 4.5 The lower and upper surfaces of a commercial frozen block of fillets; the lower is smooth and normal in appearance, the upper is white and porous showing that it is badly affected by freezer burn.

flavour after several years. In fatty fish rancidity proceeds rather faster and quality remains good for only about 6 months under the same conditions. Smoked fish and shellfish also keep well for only this somewhat shorter time. A temperature of $-30°C$ is economically realisable in practice and has for long been the recommended upper temperature for most long-term storage. To retain the full fresh colour of tuna, a temperature of $-40°C$ to $-60°C$ has been recommended. At higher temperatures, storage lives are correspondingly shorter. For example, at $-18°C$, the design temperature of most frozen display cabinets, fish fresh before freezing will keep in excellent condition for only 2–4 months. Just below the freezing-point at $-3°C$ to $-5°C$ where about 60–80% of the water is frozen out, deterioration is especially rapid and inedibility ensues after only a very few weeks. Rather interestingly at about $-1°C$ to $-2°C$ where up to about 50% of the water is frozen out, deteriorations characteristic of frozen storage proceed somewhat less rapidly than at $-3°C$ to $-5°C$ while at the same time the activities of micro-organisms are significantly inhibited with respect to completely unfrozen fish. Accordingly, with careful adjustment and control of this partial freezing to $-1°$ to $-2°C$ (a process erroneously referred to as 'superchilling') about a doubling of storage life over that at 0°C can be obtained. At the present time the cost and practical difficulty of arranging the conditions necessary render the method unattractive for many fisheries.

The storage lives of different frozen fish products at various temperatures can be summarised as in Table 4.1. The data in this table and in others of a similar kind published in specialised literature can be used to guide the producer and persons responsible for quality control in optimally planning frozen stock control and the system of cold storage (the 'cold chain'). It should be remembered that the storage life depends upon initial freshness: the staler the fish before freezing the shorter will be its life during frozen storage. Also, other factors such as packaging and ingredients and the exact criterion for assessing quality are important. Other things being equal, cooked fish stores better than raw; whole fish or fillets better store than comminuted flesh.

Because the effect of temperature is so pronounced it is of great importance in maintaining quality to keep fish close to the lowest temperature reached after freezing. Products should be transferred quickly to holding cold stores. Similarly, transfers between distribution cold stores and retail display cabinets should be expedi-

Table 4.1 Storage lives of different frozen fish products at various temperatures.

Type of commodity	Number of months in good condition at a temperature of	
	−18°C	−30°
Lean fish, whole or fillet blocks	4–8	8–24
Lean fish, individually quick frozen (IQF) fillets	3–6	6–18
Smoked lean fish	3–6	6–18
Fatty fish	3–4	6–12
Smoked fatty fish	2–4	6–12
Breaded and battered products	6–9	12–24
Crustacea	4–6	8–18
Molluscs	3–4	6–12

In all cases it is assumed that the commodity is adequately protected against dehydration. Further protection against oxidative rancidity by, for example, vacuum packaging may extent the storage lives of fatty products.

tious. Delays of over an hour at ordinary ambient temperatures are inadmissible. The temperatures of cold stores and cabinets should be checked regularly or continuously; the stored products should also be checked occasionally because under some circumstances their temperatures can be materially higher than that of the indicated store or cabinet temperature. In addition to the maintenance of a low average temperature it is important to avoid wide fluctuations since these can lead to rapid migration of moisture with resulting damaging dehydration and enhanced deterioration. All this demands a high standard of cold store, display cabinet and equipment design and management.

The signs of deterioration in still frozen fish are sometimes not particularly noticeable and are totally hidden if opaque packaging is used. Therefore poor quality stock is not self-evident and periodic checks on it to ensure that it has not been stored too long are good practice. Regular stock rotation is essential. The possibility of open or coded date-marking as an aid to stock checking should be considered.

As far as the freezing process itself is concerned it is now widespread knowledge that it has to be carried out with some rapidity: the term 'quick freezing' is in popular currency. For fish and fish products 'quick' means that the time for every part to pass between

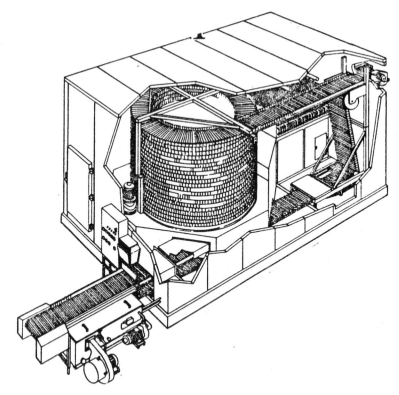

Fig. 4.6 High capacity quick freezer.

0°C and −5°C should not normally exceed 5–10 hours and that the temperature of the warmest part should be about −20°C before the product is removed from the freezing equipment. In fact, modern freezing equipment for reasons of maintaining high output can well exceed this requirement and a total freezing period of 1–2 hours should be aimed at. Some large fish like salmon and tuna can be frozen over about 20 hours and still give a satisfactory product, particularly for subsequent canning. Slow freezing over several days such as occurs when containers of wet fish are stacked in a cold store should never be allowed. If fish are kept between 0°C and −5°C microbiological spoilage can continue and, as just pointed out, frozen storage deteriorations proceed fastest in this zone. The importance of reducing the temperature to about −20°C resides in the fact that putting 'warm' fish in a cold store at, perhaps, −30°C accelerates the loss of moisture of the produce in it.

Various types of freezing equipment are available to suit different fish products. Some claim to offer advantages in terms of improved quality and yield but these advantages are generally marginal and regardless of the equipment (provided freezing is done properly) little, if any, loss in quality results from freezing. If any doubt exists as to whether produce is being frozen properly, a temperature record taken during the process should be obtained. Equipment should be likewise checked, maintained and defrosted properly.

In general, whole fish can be frozen successfully at any stage of rigor as can fillets pre- or post-rigor. However, the problems of shrinkage and surface appearance associated with fillets in rigor described for the chilled product, can also be present here because freezing merely fixes the flesh in its unsatisfactory state. The remedy has been discussed in the previous section. If there is any choice in the matter, whole round fish should be frozen pre-rigor because they tend to show least gaping when thawed and filleted.

Every time fish is frozen and thawed, even when the job is done properly, experts can detect some slight loss in quality. If a batch of

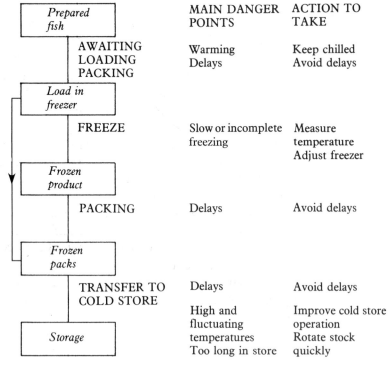

	MAIN DANGER POINTS	ACTION TO TAKE
Prepared fish		
AWAITING LOADING PACKING	Warming Delays	Keep chilled Avoid delays
Load in freezer		
FREEZE	Slow or incomplete freezing	Measure temperature Adjust freezer
Frozen product		
PACKING	Delays	Avoid delays
Frozen packs		
TRANSFER TO COLD STORE	Delays	Avoid delays
Storage	High and fluctuating temperatures Too long in store	Improve cold store operation Rotate stock quickly

Fig. 4.7 Freezing of prepared fish – simplified flow chart.

fish is frozen and thawed a number of times, these quality losses accumulate. However, up to two or three cycles, the loss is still small enough to be commercially acceptable. After that number, fluid and flavour loss and the development of toughness render the practice undesirable. Double freezing is, of course, inevitable with some accepted products, like frozen breaded or battered fillet portions and fingers made from previously frozen and thawed raw material. Double or further multiple freezing may reduce storage life somewhat over that of the single frozen product.

For the same reason that quick freezing is necessary for the maintenance of quality, so quick thawing is necessary; the general recommendations about times taken for the freezing process also apply. However, when thawing additional precautions are required connected with the fact that heat or energy has to be applied. Thawing can be accomplished by placing the frozen fish in still or moving air, immersing in still or moving water or by spraying with water, but in no case should the temperature of the medium exceed 20°C otherwise there is a danger of the outer layers spoiling, becoming unduly softened or waterlogged and losing flavour before the interior has thawed out.

Transfer of heat by condensing water on the fish in a vacuum chamber is a new development that offers the advantage that the surface temperature can never rise above a low value. On the other hand, it has the disadvantage that the desirable fresh red colour of tuna is partially decolourised when subjected to the vacuum. Water thawing of fillets is not recommended because of dangers of waterlogging and loss of flavour. If air thawing is employed precautions against the fish drying out must be taken. The air in machines for blast thawing blocks of frozen fish on an industrial scale is for this reason humidified with water sprays. Energy for thawing can be supplied by passing electricity or high frequency radiation through the fish in special equipment. In this less commonly used method the fish warm up more or less uniformly. Quality can suffer here as a result of improper use leading to overheating and partial cooking of the fish. Careful control of the equipment is indicated.

If thawed stocks have to be stored it is an advantage to halt a little before the thawing is complete so leaving a protective reserve of cold in the product. Slightly warm thawed fish if not processed immediately should be chilled with ice. As well as maintaining freshness chilling also aids cutting and the attainment of good yields.

Fig. 4.8 The use of mechanical air blast thawers as illustrated here and of other types has brought under control one problem of handling efficiently large quantities of frozen fish.

Controlling the deteriorations under discussion by means of reducing and stabilising the temperature does not exhaust the possibilities because further means of reducing both dehydration and oxidation are available.

The first can be achieved by reducing or preventing the loss of moisture through either coating the frozen fish or product in a layer of ice or placing it in a more or less tightly fitting wrapper. The layer of ice is deposited by simply dipping the frozen product in or spraying or brushing it with clean water. This process, known as 'glazing', is a cheap and effective remedy of wide application. Under drying conditions in a cold store the glaze will evaporate rather than the moisture in the product. Thus, it must totally cover the product and be of sufficient thickness to provide protection over a reasonable period. Thicknesses of 0.5–2 mm, representing for most products a weight gain of 5–15% of applied water, are in general sufficient. If the glaze evaporates completely from any part of the product before the end of the storage period it should be renewed. Products frozen as separate pieces ('individually quick frozen' – IQF) can be glazed as such, but blocks of whole fish or fillets are normally only glazed on the outside after freezing is completed. Better protection of small pelagic species is afforded if the spaces between the individual fish are totally filled with ice. By freezing such fish and water together in a plastic or otherwise waterproof bag fitting within a vertical-plate freezer, additional advantage in terms of protection against physical damage to these vulnerable species is gained. If it is physically impossible to glaze, some form of wrapping is essential.

The materials and styles of packaging are numerous and vary from the flexible plastic polybag for retail sale of products like shrimps to waxed paper or polythene-lined board for industrial blocks of fillets. Their efficacy depends on their degree of imperviousness to water vapour at low temperatures and the degree to which the product is closely and completely enclosed. However, even under the best conditions some slight moisture loss over long periods has to be accepted because water tends to evaporate from the product surface and condense on the interior of the package to form a snow ('in-package dessication'). Blocks of frozen fish on pallets can be usefully protected from dehydration by enclosing the stack in a fairly tight-fitting cover. Coatings such as batter also afford a useful measure of protection against dehydration.

Oxidative rancidity can be controlled by preventing oxygen from

Fig. 4.9 High quality herring can be obtained by freezing at sea; in this operation bags held in the spaces of a vertical plate freezer are filled with fish and sea-water. After freezing and release from the freezer the blocks appear as in Figure 4.10 (below), with the fish well protected by ice and the outer bag.

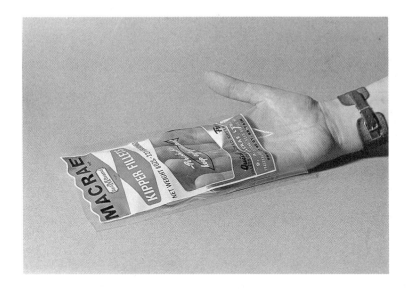

Fig. 4.11–4.12 Two types of protective and attractive packs that are available for frozen fish.

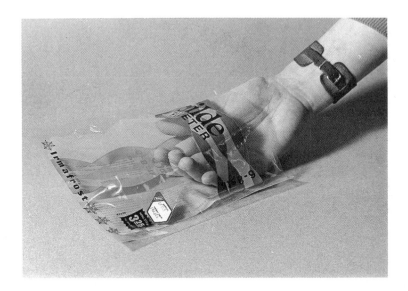

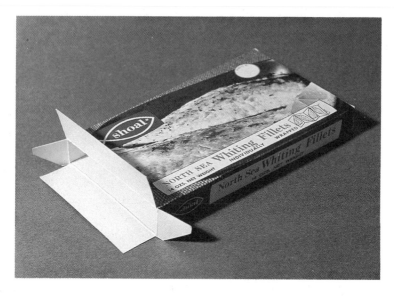

Fig. 4.13–4.14 Other types of packs for frozen fish that are both protective and attractive.

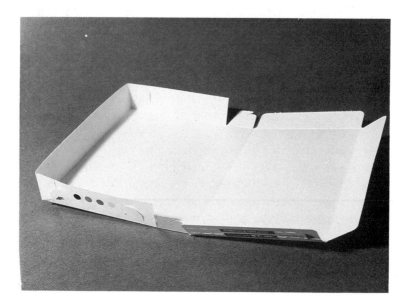

attacking the unsaturated lipids in the fish. The means are the same as those used in preventing dehydration, namely interposing a physical barrier, in this case between the product and the oxygen in the air. The glazes used to protect against dehydration are, in fact, also useful in protecting against oxidation and a doubling of storage life of fatty species is typically obtained by thorough glazing. Packaging in oxygen-impermeable flexible films also offers practical extension of storage life but here the oxygen in the package must be removed before storage either by evacuating the space between film and product or by replacing the air with an inert gas like nitrogen. Vacuum packaging is found to be more practical, though it is only applied to expensive, small lots, like shrimps and trout. Packaging in rigid plastic or metal containers is too expensive. Certain chemicals known as antioxidants have the property of inhibiting oxidative rancidity but they have to be mixed intimately with the lipid in the fish; in general this is not possible so they have very limited application to fish and fish products. Salt (sodium chloride) has a pro-oxidant action and as far as possible should not be mixed with or applied to fish destined for frozen storage. Partly for this reason smoked fish (which are also brined during preparation) do not keep as well as unsmoked under the same frozen storage conditions. Given the choice, fish should be brined and smoked after frozen storage and thawing rather than stored in the brined and smoked state.

Up to the present time it has been found impossible to alleviate by means other than control of temperature, the textural deterioration caused by protein denaturation. Application of polyphosphates as described earlier can offer by a kind of sealing mechanism some measure of protection against fluid loss of thawing, but this is no radical cure of the underlying change. Similarly, problems of frozen storage of Crustacea can be avoided by cooking the animals before freezing: if frozen raw, lobsters for example, tend to deteriorate rapidly and are difficult to shell after thawing; cooking shrimps before freezing also facilitates peeling post-thawing. Since fish of low pH tend to deteriorate faster than those of high pH, selection before freezing and frozen storage offers one means of controlling the situation. However, the application in practice of this idea presents formidable difficulties. In general, it will not be known in advance which fish in a catch will have low pH and it would be necessary to test and possible segregate large numbers of fish in a batch before freezing. Rapid and effective means of doing this, even if it were economically justified, are not commercially available apart from application to exceptionally large fish like tuna.

Minced or comminuted fish is often frozen into blocks for further processing. Also, small proportions of minced fish are sometimes added to fillets before the mixture is frozen into the blocks subsequently used to prepare fish fingers or portions. In recent years manufacture of the product called surimi, once confined to Japan, has been increasing in other countries. Surimi is minced, washed and stabilised fish: as stated in the Preface, it is too specialised a product to be dealt with here. The mince is sometimes recovered from fish or parts of fish which were formerly used only for the preparation of fish meal. A good deal of edible flesh is, for example, attached to trimmings on the skeleton after the fillets have been removed. Recovery is effected by passing these materials through special machines called deboners. In order to obtain good quality mince it is very necessary to choose the materials carefully. Parts containing much blood, portions of viscera or pigments must be avoided, otherwise the recovered mince will be contaminated. It has been found that mince containing blood or particularly tissue from kidney or liver deteriorates in flavour and texture very rapidly when frozen. In addition there is evidence that the mince from some species is particularly prone to this type of deterioration. Some amelioration of these changes can be achieved by limited washing of the mince before freezing.

As noted already flavouring agents are occasionally added to frozen fish, but this practice is cosmetic rather than real quality control.

(ii) Defects of freezing, storage and thawing

General points relating to damage, excessive dehydration and taints have been dealt with already. Cross-tainting during distribution and storage of frozen fish by strongly odoriferous commodities like oranges can occur. Remedies are proper segregation and, if possible, adequate ventilation of the storage space.

As far as fish frozen at sea in the form of blocks of whole fish or fillets is concerned, the points requiring attention additional to those already covered are the integrity of the blocks (they should cohere and not fall apart on handling), inadequate glazing or wrapping, incorrect weight or size of block. The species or size of fish or

Fig. 4.15 A poor quality block of frozen fillets ('laminated block') showing voids caused by underfilling the mould before freezing; such a block would be quite unsuitable for further processing into fingers or portions.

pieces within blocks should be as uniform as possible, though there are severe limitations on the amount of pre-freezing grading that can be achieved at sea. In fish frozen at sea individually in brine (tuna, shrimp) excessive salt penetration through too long a period in the freezing medium should be avoided.

Blocks of fillets are often designated as skin-on, skin-off or boneless; the contents of such blocks should conform to what is stated. So-called laminated blocks of fillets for further industrial processing into cut-up portions and fingers are usually subject to detailed specifications. Requirements of minimum incidence of bone, skin, scales, connective-tissue, belly membrane, etc. should be observed. Departures from specified size and geometry should be avoided; considerable wastage occurs in laminated blocks that are not uniform in size and shape. Edges and corners should be sharp and at correct angles, faces smooth. Voids and regions of pure ice (caused by excessive water in the spaces between the pieces of fish) are serious defects. Colour should be uniform and for many purposes, as light as possible; streaks of dark-coloured flesh are generally undesirable. The proportion of mince or small fillet pieces in laminated blocks should not exceed specification. Systematic visual inspection combined with feeling for bones with the fingers

Fig. 4.16 Good quality frozen block with sharp edges, flat faces and almost complete absence of voids.

during the making of blocks is the best primary method of control. End product inspection of samples from production batches according to a plan offers a good additional safeguard.

Fish or pieces of fish meant to be individually quick frozen should be easily separable at the time they are used and not adhering in lumps. The proportion of glaze should not be excessive otherwise complaints about underweight may arise. In some countries the amount of water added to fish products during processing is regulated: steps should be taken to ensure that the limits are adhered to. IQF fillets and shellfish are often required to be graded for size, thus the range of permissible variation should not be exceeded.

The special defects encountered with the large range of battered and breaded frozen products have been dealt with in the section on chilled products. Likewise, contamination with shell and other unwanted materials in the case of shellfish has been covered. It should be noted that in all pre-prepared frozen fish dishes the quality and quantity of the ingredients also need controlling. The correct degree of cooking for pre-cooked fish dishes should be carefully determined and rigidly maintained. Tainting from pouch materials is possible in the case of boil-in-the-bag products; only tested and recommended films or laminates should be used. Defects arising from overtreatment with or illegal use of chemical additives or flavouring agents are similar to those already enumerated. There has been a recent move away from the use of additives such as polyphosphates in frozen fish products. Frozen fish can, in addition, be contaminated with refrigerants such as ammonia or trichloroethylene. Although rapidly diminishing in use, some fish is still frozen by direct immersion in the liquid or vapour of a purified fluorocarbon (dichlorodifluoromethane). At normal temperatures and on cooking most of the refrigerant absorbed inevitably by the product evaporates but traces always remain. Some countries permit products frozen in this way and containing such traces, others currently do not.

3 Smoked fish

The process of smoking and the brining that usually precedes it cannot entirely mask or disguise spoilage or defects that are present in the raw material. Therefore, the quality of smoked fish depends to a large degree on that of the raw material. Nevertheless, the strong flavours introduced by salt and smoke constituents are capable of concealing some incipient spoilage flavours and it is quite possible to make acceptable smoke products from fish that in the raw state might be judged unacceptable. The rise in temperature undergone during smoking may also volatilise some of the objectionable compounds present in spoiling fish. The characteristic yellow, golden brown or brown colours imparted to the fish by smoke constituents or artificial smoke colours are capable of disguising minor imperfections of colour in the fish itself. Sometimes yellow or brown dyes are used before smoking to impart an attractive deeper colour than is obtainable by normal smoking alone. If spoiled fish is used there is a danger the food dye may lose its colour or even turn pink, with obvious disastrous effects on quality. The change in colour is caused by the high pH of the spoiled fish.

Apart from freshness, the quality of smoked pelagic fish depends to a considerable degree on the initial fat content. Samples having low fat of say 1%, tend to taste drier and less succulent than those of high fat content. Moreover, the former are thinner and have a less glossy, attractive appearance than the latter.

(i) Deteriorations

The spoilage of smoked fish at chill temperatures is basically similar to that of unsmoked fish, and the same pattern of odour and flavour changes is discernible. Some differences may exist due to the fact that the intrinsic and microbiological enzyme changes are modified to some extent by the presence of salt and smoke. The partial or complete cooking given in the so-called hot smoking process also modifies the spoilage pattern, though the end character is still ammoniacal, faecal and sulphide. The low moisture content of smoked fish favours the growth of moulds that originate in the wood used for smoking; the product, especially after wrapping, may therefore suffer additional musty-flavoured deterioration from these organisms. Visibly mouldy fish should, of course, be thrown out. On storage the fresh aromatic smoke flavour of newly cured fish becomes weaker, blander or unpleasantly tarry in nature. Rancidity is much more noticeable in stale smoked fatty fish than in unsmoked.

The combined effects of salt, smoke constituents (some of which are bacteriocidal) and the drying that accompanies the smoking process are significantly preservative and the storage life of smoked fish is always longer than that of the unsmoked article held under the same conditions. The precise extension of storage life depends on the vigour of the treatment; more salt, smoke or drying all tend to increase storage life. Heavily brined, smoked and dried fish will remain edible for several weeks at 0°C and for several days at 15°C–20°C. The modern tendency is to reduce reliance on heavy curing as a means of preservation and to chill (or freeze) the lightly smoke-flavoured product. The main factor in controlling deterioration in stored smoked fish is, again, temperature though in this case the direct application of ice to the product is inappropriate. Fish after smoking is warm and should be cooled in cold air as rapidly as possible to avoid unnecessary spoilage. Packing warm products closely can lead to 'sweating', that is the tendency of moisture to come to the surface and render it sticky.

(ii) Defects

The production of smoked fish of good and consistent quality basically depends upon maintaining certain levels of moisture, fat,

Fig. 4.17 Where capital and facilities are available the smoking operation can be efficiently and reproducibly controlled by the use of a modern mechanical smoking kiln such as this.

salt and occasionally dye content and of degree of smoke deposition. Any major departure from these means a defective product. For example, a product that has been excessively dried will be defectively firm, case-hardened or dry to the palate; an overbrined product will be too salty; an overdyed product will not have the correct colour. Not only must these requirements be correct they must be uniform for each batch of product. All this implies good control of the production process. Production control is considerably simplified when modern mechanical forced draught kilns are used because temperature, humidity and air flow over the fish can be varied easily to suit different conditions or can be set to predetermined values. Conditions in well-designed mechanical kilns are much more uniform than in natural draught kilns. The brining process can also be brought under control through the use of equipment in which the fish is transported through a batch of nearly saturated salt solution.

A good deal of fish in developing countries is still smoked and partially dried by simply hanging it over an open fire. This is very difficult to control and some of the product is inevitably wasted either during smoking or by subsequent spoilage. Furthermore, valuable fuel and wood are used very inefficiently. Simple, cheap kilns made out of locally available materials have been devised which, although not ideal, can help to greatly improve this practice.

The appearance of smoked fish adds considerably to its appeal and particular care should be taken with gutting, splitting, filleting or trimming preparatory to smoking. A good quality product should possess a smooth, glossy pellicle; a dull, ragged or gaping surface severely detracts from appearance. The production of a good gloss depends upon correct brining, adequate draining after brining and correct drying in the smoking kiln. Stale, poorly cold-stored, poorly cleaned and washed or pre-rigor fish give poor gloss. Overbrining may result in an undesirable white crust of salt crystals on the surface of the product. Inadequate control of the wood fires in traditional types of smoking kiln gives rise to a number of defects including poor drying and a sticky surface, a disfiguring deposit of smuts, singeing and overheating resulting in a semi-cooked split skin. Dirt or coagulated protein on the surface of the product may be picked up from the brine bath which accordingly should always be kept clean and free of debris.

In order to obtain a satisfactory smoke flavour the correct un-contaminated wood or other cellulosic material must be used; some woods impart unpleasant resinous or acrid flavours. Checks should be made to ensure that the wood blocks, chips or sawdust have not been treated with preservatives and do not contain glues or rem-nants of plastics. It is possible to avoid the use of the smoking process altogether by adding smoke-flavouring mixtures to the fish. It some cases it may still be necessary or desirable after the application to reduce the water content of the fish by a drying stage. Such flavouring agents have been much studied in the past but their practical use is growing only slowly. Peculiarly flavoured products can arise from their misuse; the concentration must be strictly controlled.

4 Canned, bottled and similar products

The preservative action of most canning and similar processes depends upon the heat inactivation of intrinsic and microbiological enzymes and the protection of the product from subsequent attack by micro-organisms and atmospheric oxygen. With a minority of these products which will not be dealt with here, preservation is achieved not through the application of heat but by the action of high concentrations of salt or acid. The hermetically sealed container used also protects against damage and contamination with dirt. Control of quality again depends primarily on selection of good quality raw material. Major physical damage, autolysis, belly-burst, discoloration, off-odours, and contamination can all irrevocably ruin the raw material for canning.

Insofar as canning and related processes are meant to render batches of product virtually sterile – and do it consistently – it is of considerable advantage to avoid raw material of very high bacterial load. Otherwise, sterilisation procedures found effective with ordinary material may yield an unacceptably high proportion of non-sterile packs. On the other hand, the canning process may sometimes give highly acceptable products from raw material that possesses slight, incipient or frank spoilage features. That is not to say, of course, that once the raw material quality necessary to produce the finished article has been determined, something lower can be accepted with impunity. A related point is the fact that fish can be too fresh: in rigor it is often difficult or impossible to handle in the stages preparatory to canning; very fresh tuna is more liable than less fresh, to show 'scorching' (darkening of the meat surface exposed to the headspace).

Since the size and shape of containers is fixed for any given finished product, the size of the fish usually has to be controlled within close limits if the 'fill' and appearance after opening are to be right. If, as is usual, the exact uniform size of catch cannot be selected, mechanical or hand grading from mixed lots in the factory is necessary. Fatty, pelagic species are by far the most commonly canned and their fat content often determines the quality of the end product. For example, to make the best quality canned sprats and sardines the fat content of the raw material should lie between 7 and 15%. By selecting batches of fish from known grounds and at the proper seasons, fish having a fat content within the required limits can be expected, though the value might need to be checked by chemical determinations. Some small species like sardines and sprats are canned with the visceral organs virtually intact. In these cases it is advantageous or even a necessity to process fish having an empty gut ('clean' fish). This condition can only be obtained naturally by catching the fish when they are not feeding or by storing them alive (impounding) until the gut empties. Unclean fish in this sense are prone to belly-burst and a disfiguring oozing of gut contents during drying or retorting results.

Further quality defects in canned fish that are dependent upon compositional differences are 'greening', colour variations and 'struvite' formation.

The first is a grey or greenish-grey colour in tuna that detracts from the normal buff or pinkish-tan colour. The cause is not entirely elucidated but is connected with changes in the haem (blood and related) pigments: thorough bleeding and hence reduction in the concentration of haem pigments after catching tends to reduce the incidence. Although other measures have been proposed none are yet universally acknowledged to be effective.

Colour is a major compositional factor influencing consumer acceptance in canned salmon and tuna. The shade must be characteristic of the species traditionally canned and as uniform as possible within and between batches of finished product. This is achieved by visual inspection of the raw material by trained personnel and selection for canning of those samples having the correct colour characteristics.

'Struvite' is a hard, glassy crystal of magnesium ammonium phosphate formed from natural constituents of the fish in the period after heat processing. The occurrence of the crystals because they look just like glass causes very adverse consumer reaction. Struvite is of fairly wide occurrence in canned fish. The incidence in tuna, especially albacore (*Thunnus alalunga*), is associated with high pH. Measurement of this parameter on the raw material and selection for canning in the solid form of fish having pH values below 6.0 eliminates the defect. Fish having higher pH values can be mixed in suitably small pieces with low pH fish without struvite forming. Struvite in other canned species can be reduced by either lowering the pH of the contents with citric acid before canning or adding about 0.5% sodium hexametaphosphate or sodium acid pyrophosphate.

Preparatory cutting and cleaning operations should be carried out neatly, thoroughly and in accordance with the requirements of the pack. These include: butchering (heading and eviscerating) of tuna, followed by pre-cooking and then removal of skin, bones and red or brown muscle ('blood meat') after cooking and cooling; 'nobbing' of raw herring (removal of head and gut in one operation) and ensuring emptiness of body cavity; trimming of head and tail of sprats and sardines after drying, smoking or partial cooking; shelling and shucking of shellfish; de-veining of shrimp or prawn. Machines for these operations should be kept well adjusted and their cutting devices sharp. Skin, scales, fins, blood, blood-clots and other debris should be removed or washed off as appropriate before filling into cans or containers. The latter may themselves need inspecting for cleanliness or washing; damaged or defective containers must be rejected. In order to avoid impairment of the appearance of the product by liquor cooked out during retorting, it is often necessary to remove a portion of the water in the tissue beforehand.

The procedure needed to accomplish this should be carried out in a controlled fashion which ensures that either the correct moisture content is attained or all the protein in the fish is coagulated before canning. Pre-adjustment of the raw material temperature and care with temperature and time of cooking are points to note. Methods include steaming in the case of tuna, herring, sprat and some sardines; frying of other types of sardine; smoking of sprat (brisling); heating in brine in the case of other types of tuna. If the pre-cooking is conducted in the final container the cooked out liquor must be drained adequately – normally by inverting the container.

The fish should be arranged neatly in the manner traditional for the pack in question and fit compactly; spaces allow the fish to move about with consequent risk of break-up. To meet the requirements of legislation relating to fair trading practices and of customers' needs, close control of the weight of fish and the proportion of fluid packing medium is very necessary. Some regulations specify the drained weight that should be packed. Care should be taken that the numbers or proportion of small or marginally inferior pieces of fish are not excessive. Species are only rarely allowed to be mixed in the same container or batch; similarly, checks that the correct style (for example, chunk or shredded in the case of tuna) is being packed are needed. The quality of non-fish ingredients – salt, flavouring agents, acidulants, additives, oil, sauces, vegetables – all require checking by the processor or guaranteeing by the ingredient supplier.

After filling it is essential to form a partial vacuum in all but the smallest cans. If this is not done the gas in the spaces between product and internal surface will expand and may cause the can to bulge during heat processing and subsequent cooling; removing oxygen also helps to prevent oxidative rancidity developing during storage. Exhausting, as the process of creating a partial vacuum is called, is critical to the success of the canning process and it is

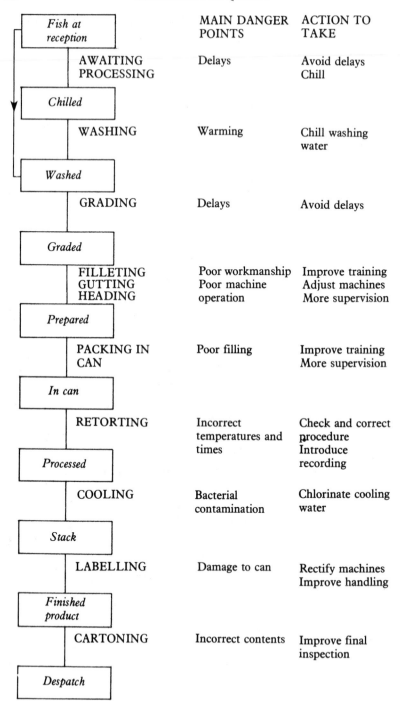

	MAIN DANGER POINTS	ACTION TO TAKE
Fish at reception		
AWAITING PROCESSING	Delays	Avoid delays Chill
Chilled		
WASHING	Warming	Chill washing water
Washed		
GRADING	Delays	Avoid delays
Graded		
FILLETING GUTTING HEADING	Poor workmanship Poor machine operation	Improve training Adjust machines More supervision
Prepared		
PACKING IN CAN	Poor filling	Improve training More supervision
In can		
RETORTING	Incorrect temperatures and times	Check and correct procedure Introduce recording
Processed		
COOLING	Bacterial contamination	Chlorinate cooling water
Stack		
LABELLING	Damage to can	Rectify machines Improve handling
Finished product		
CARTONING	Incorrect contents	Improve final inspection
Despatch		

Fig. 4.18 Canning – simplified flow chart.

essential to carry out the methodology strictly according to the procedure worked out beforehand for the product and type of can. A lower partial vacuum is in general required for cans expected to be exposed during storage to high ambient temperatures or very low pressures. Securing the lid to the body of the can by seaming is carried out after or as part of exhausting. Again, this is a critical operation the efficiency of which needs to be checked periodically on the closed cans and on the seaming machine. The various points involved in an examination of can seams and of deformed cans are described in specialist literature. Similarly, the effectiveness of closures used on bottles and other types of container needs frequent checking. After closure, containers may need to be thoroughly washed to remove adhering debris that would otherwise dry hard on the outside during the following retorting; periodic checks on the efficiency of this step are indicated. To reduce the risk of spoilage at ambient temperature, delays after filling and before retorting should be kept to a minimum.

Correct retorting times, temperatures and procedures to suit different sizes and shapes of container, different products and different types of retort are available in specialist literature. The primary objective in all cases is to ensure that the most heat-resistant micro-organisms in the pack are killed. A variety of such organisms may be present, one of the most important being the spore of *Clostridium botulinum* since, as described later, it presents potentially a risk of lethal food poisoning. In practice it is not possible to achieve 100% sterility in all containers without excessively damaging the product and the aim is to reduce the risk of having pathogens present to an extremely low level whilst producing a food that will keep for a long time at ordinary temperatures without spoiling. Great care, therefore, should be taken to follow the laid-down prescription meticulously; time and temperature records should be made, preferably on automatic instruments. Cooling after heat processing should be carried out rapidly to avoid over-processing and to reduce the risk of struvite formation. A form of over-processing ('stack burn') occurs if cans are packed while still hot into outer cartons or closely together in piles. Under these conditions cooling can be very slow. The remedy is to cool thoroughly before packing or stacking. The special methods of pressure cooling normally included in retort operation also require periodic checking. Potable quality water should be used for cooling and washing. Traces of water are prone to be sucked into the can during these stages and if micro-organisms are so introduced spoilage may ensue; chlorination of plant water helps to eliminate this possibility by very greatly reducing the numbers of micro-organisms.

The seals in containers can be weakened and broken by rough treatment and contaminating micro-organisms gain access to the contents. Thus care in handling filled containers is necessary at all times. A well-publicised case involving two deaths and severe injury to two other people caused by *Clostridium botulinum* poisoning from canned salmon dramatically illustrates this point. It is almost certain that after heat processing the micro-organisms entered the contaminated can through a tiny hole in the end seam which was damaged in the factory.

It is good practice to emboss cans or mark containers with a code showing a few essential details of production so that defective batches may be readily identified, isolated and if necessary removed from sale. Coding also makes it possible in principle to trace back the cause of defectiveness.

(i) Deterioration of canned and related products

These are of two kinds; microbiological and chemical. The first arises either from insufficient heat processing during retorting or from contamination through seals at any stage up to detection. In both cases the result is almost invariably a foul-smelling product and often a swelling (belled-out end pieces) or bursting of the can due to microbiological generation of gas. More rarely, as in the case of goods contaminated with *Clostridium botulinum*, frank spoilage may not be very evident. Also, rarely in the case of fish products, microbiological spoilage may not result in the generation of gas and development of pressure. Various kinds of can deformity with special names (springer, hard swell, etc.) are caused by varying degrees of spoilage.

Chemical spoilage arises typically from the gradual attack of substances in the product on the metal of the can. Commonly these substances are the acids in sauces or packing medium, the result being internal corrosion, generation of hydrogen with swelling of the can and discoloration of the contents. In addition, traces of metal are leached from the tinplate, contaminating the product and so possibly causing a health hazard. This form of deterioration is virtually eliminated by coating the interior of the can with a protective lacquer or by using aluminium cans. Only in very rare cases do canned fish products contain dangerous concentrations of metals.

A second form of chemical spoilage which does not affect the fish itself is internal blackening of tinplate cans. Volatile sulphur compounds derived principally from the fish (especially Crustacea) can gradually react under some circumstances with metals in the plate to form a black or dark disfiguring deposit of iron sulphide or tin sulphide. Much of the problem can be overcome by using tinplate coated with special lacquers including one containing zinc oxide which still reacts with sulphides but in this case to give a white unnoticeable compound. Further protection is afforded by interposing parchment paper between the fish and the can interior. Aluminium cans do not suffer from this problem. A similar phenomenon sporadically affects the packed meat itself, particularly with canned shrimp or prawn, and crab. Here the sulphur compounds react with iron and probably also copper present in the meat. The metals can be either of natural occurrence or picked up adventitiously during preparation. The addition of small amounts of citric acid, citrate or phosphoric acid to the final filling medium or dipping of the shellfish in solutions of these substances before canning both inhibit the release of sulphur compounds and act as chelating agents for the metals, so considerably alleviating the problem. More acid needs to be added to staler raw material. For unknown reasons, a small number of species even when canned fresh do not respond to this treatment but success can be achieved by adding to the pack a very small quantity of another much more effective chelating agent, disodium ethylene diaminetetraacetic acid.

Both microbiological and chemical spoilage often take some time to manifest themselves and hence it is of advantage to store cans for a few weeks or months before final inspection and despatch. Storage also allows some products to go through a desirable maturation.

Some fish products packed in glass jars or bottles become bleached or discoloured after prolonged exposure to sunlight; protection with suitable outer packaging and a short period on display are obvious remedies.

(ii) Defects

With a complex process like canning including many preparation stages, it is not always possible to make a clear cut distinction between deteriorations occurring over a period and defects. Many of the faults arising during preparation and processing if not corrected can obviously give rise to defects. These include: off-odours, off-tastes and discolorations resulting from the use of spoiled raw material; neatness of trimming and packing; damage to cans; presence of excessive amounts of shell, fins, skin; use of incorrect ingredients or additives. Other defects not so far included are the presence of undue amounts of fish debris in the packing medium; poor appearance, consistency or colour of the packing medium; incomplete softening of bones; adhesion of skin to container or of fish to one another; excessive loss of scales in species (like high quality sardines) intended to be packed scales-on; excessive fissures in the flesh or in the ventral region; incomplete coverage of fish by packing medium. The correct weight of fill and of drained weight of fish where these are requirements must be complied with, taking into account allowable tolerances. Similarly, the species, size or count (in the case of shellfish) of the contents

Fig. 4.19 An attractive, neat arrangement of fish in the can as in the upper two examples of sprats is essential in good quality packs; the two lower commercial packs show a number of defects including jumbled arrangement of fish; presence of fish fragments, partial loss of skin and tearing of the opened tinplate lid.

should meet the requirements of the customer and agree with statements on the label. In packs containing chunks, flakes or shreds of fish, the particle size should be within the permitted limits. A defect that occurs sporadically and unexpectedly in skipjack tuna (*Katsuwonus pelamis*) and albacore is an orange discoloration. Like 'greening' this is not detectable in the raw material and the cause is unknown. End product inspection is the only way of detecting its occurrence.

When badly frozen and thawed salmon and mackerel are used as raw material for canning, the finished product is sometimes rendered unattractive in appearance by the formation of clumps or layers of coagulated protein ('curd'). The condition is alleviated by dipping the fish before canning in brine or a solution of tartaric acid which seals the cut surfaces and prevents the escape of fluid containing potentially coagulable protein. Alternatively and much more effectively, the upper surface of the fish in the can before retorting is treated with a small quantity of a proteolytic enzyme similar to a meat tenderiser. During the initial stages of heating the enzyme breaks down the protein on the surface that might otherwise form curd. Further heating destroys the enzyme and so avoids wholesale liquification of the can contents. The result is a clean appearance on opening the can.

A grey or bluish discoloration arising from blood pigments and possibly the presence of iron occurs principally in the relatively small amount of canned horse hair crab (*Erimacrus isenbeckii*) and blue crab (*Callinectes sapidus*). The defect can be at least partially overcome either by cooking at a moderate temperature which coagulates the meat protein but allows the still fluid blood pigments to drain away or by treating the meat before cooking with brine containing a small amount of aluminium sulphate or ethylene diaminetetraacetic acid.

Corrosion may affect the exterior of tinplate cans or metal fastenings. Condensation of moisture or inadequate washing are the main causes and should be guarded against, the first by provision of adequate ventilation and an evenly controlled temperature of storage.

5 Salted fish

Preservation in these products is accomplished by the inhibiting action on micro-organisms and enzymes of both high salt concentration and the considerable dehydration that accompanies processing. Success in making a good quality product invariably depends upon achieving in the early stages a rapid enough increase in salt concentration and in concomitant dehydration to prevent spoilage. Salting is a fairly slow process often conducted at ambient temperature. Salt penetration occurs from the outer layers so that for some period the inside of the fish will remain unsalted and of normal water content. If neither salt penetration or rate of dehydration are sufficiently rapid, micro-organisms on the inside will multiply and spoil the fish. To this end the salt needs to be brought into as intimate contact as possible with the fish flesh and if at all possible the fish should be kept cool throughout the early critical stages of salting. These principles have been known for centuries and are embodied in traditional methods such as splitting large fish to give an enlarged surface area from which the salt can diffuse to the centre of the flesh, and pressing in stacks to aid rapid removal of water.

Although it is necessary to have very fresh raw material, if the choicest salt fish (especially lightly salted) is aimed for, it is undoubtedly true that for the manufacture of many kinds of salted product possessing characteristically ripe odours and flavours, quite stale fish unsuitable for direct human consumption may still be suitable. In fact, it is common practice on some markets for stale fish to be bought expressly for salting. Gross physical damage, contamination with dirt or excessive autolysis like belly-burst in pelagic species will, of course, render fish unsuitable. Very fresh fish is required for some special lightly salted products like the Gaspé cure because otherwise spoilage would set in before salting and drying had been completed. In order to make the palest coloured product required for some markets, fish in Scandinavian countries is thoroughly bled immediately after capture; indeed the practice is regulated by official rules. All types of white fish can be dry salt cured, though elasmobranchs may give trouble with evolution of ammonia or amines unless the salting and drying processes are carried out quickly. Pelagic fish of different fat contents can be pickle-cured successfully but those with the highest fat content make the best products.

The quality and type of the other raw material – salt – is a matter of some importance. It should be of reasonably small grain size to facilitate close contact with the fish surfaces and rapid dissolution, but not so fine as to impede drainage of expelled flesh juices. Salt containing more than traces of iron or copper gives rise to unsightly yellowish or brownish colour in finished white fish products; it should be avoided. On the other hand it should contain about 0.5% calcium plus magnesium (as sulphates) because these metals impart a desirable whiteness and rigidity; high concentrations are undesirable because they cause excessive bitterness and brittleness. White fish cured with pure sodium chloride tend to be flexible and amber in colour.

Success in producing salt fish of consistent quality depends upon closely following traditional methods paying particular attention to the following points: the ratios of salt and fish that are required for end products possessing the correct degree of cure (salt and moisture content); adjustment of this ratio to take account of different sizes and thicknesses of fish; the method of splitting or gutting the fish; the method of stacking and restacking in the case of white fish or of packing in barrels or vats in the case of pelagic species; the ambient temperature; the method and degree of drying after salting. Gutting, heading and splitting of large round white fish should be done neatly and the appropriate amount of backbone removed from the head end. In some cured herring the gills and part of the viscera are removed in one operation ('gibbing') leaving the pyloric caeca and gonads (reproductive organs) behind. The presence of the enzymes provided by the former is essential for the formation of the correct ripe flavour.

The aim in the salting stage is to obtain an even distribution of salt through the fish and therefore uniform sizes of fish should be stacked together or placed in the same region of the barrel or pickle container; also more solid salt should be placed near the thickest part of the fish. Air spaces should be avoided in order to reduce the risk of rancidity developing. In this connection, barrels or wooden containers used for holding fish and salt should be as airtight as possible; the entrance of air can cause the development of rancidity. After stack ('kench') salting has been completed, which takes 1–3 weeks, and the fish then immediately dried, the surface often carries a white, rough deposit of salt crystals. For some markets this appearance is undesirable and an intermediate step is introduced which consists of briefly washing off the salt and re-stacking in order to smooth the surface of the finished dried article.

Drying should be carried out as swiftly as possible to reduce the risk of spoilage. The humidity of the air needs to be low and the temperature as high as possible consistent with the avoidance of cooking and early case-hardening. Good control and rapidity of drying in humid, cool climates can be obtained by suspending the salted fish in a chamber through which warm air is passed. Salt fish can be dried out of doors in dry, warm climates but protection against rain, insects, birds and strong direct sunlight is necessary.

Even and more rapid drying throughout individual fish and between the fish in a batch is aided by re-stacking ('press-piling') for several days after some drying has taken place. This has the effect of equilibrating the moisture content throughout the mass; afterwards drying is resumed.

(i) Deteriorations

Apart from the deleterious effects of certain metals just discussed, the only other form of deterioration occurring during processing is microbiological. Because the anaerobic conditions in pickle or wet curing of pelagic species rarely support the growth of microorganisms (assuming of course that the salt is applied properly) these deteriorations are associated almost entirely with dry or stack curing of white fish. In uncommon types of pickling of whole herring using mixtures of salt and sucrose, the process is occasionally troubled by the occurrence of 'ropy-brine' – a slimy condition caused by the bacterial formation of a polysaccharide (levan) from the sugar. The trouble can be avoided by keeping the pickle chilled.

Four types of microbiological deterioration of salted white fish are distinguishable: slime, 'putty fish', 'pink', and 'dun'. During conditions of high ambient temperatures, high humidity or inadequate penetration of salt, certain groups of bacteria capable of living in low concentrations of salt (6–12%) may multiply and attack the fish, resulting in a sticky, slimy coating and off-odour. If the condition is not too far advanced, some at least of the fish in the stack may be salvaged. The remedy is to reduce if possible the temperature and humidity, to accelerate drying of the outer layers by increasing ventilation and to enhance penetration of salt. The second, somewhat related deterioration, occurs in the thick parts of the fish where the rate of increase in salt concentration is slowest. Again, degradative bacteria may gain a hold if the temperature is too high and reduce the fish to the consistency of putty.

'Pink' takes the form of spots or areas of pink or reddish coloration on the outside of the stack. If left untreated the condition will spread to cover and partially penetrate the whole stack, turning the fish soft and evil-smelling in the process. It only occurs when the salt concentration is high (over 10–15%) and is caused by the growth of certain groups of pink or reddish coloured bacteria (halophiles) that have a requirement for such conditions. They originate in some kinds of salt and, assuming the premises are not contaminated, the occurrence of the deterioration can be prevented or reduced by changing to salt more or less free of the offending bacteria, by heating the salt before use, or since the contamination tends to decrease with time, by using salt that has been stored for a year or so. Alternatively, the fish in the stack should be kept as cool as possible. Premises and equipment can harbour halophiles and if an outbreak of 'pink' occurs a thorough cleaning programme should be instituted using plentiful amounts of water preferably containing a disinfectant suitable for food use. Fumigation with sulphur dioxide is also useful.

'Dun' is the term used to describe a peppering of light brown or fawn spots, particularly noticeable on the cut surfaces. The causative agent is a mould (more than one kind is probably implicated) that is able to grow at salt concentrations of up to 10–15%, and in contrast to 'pink' is more often seen on light cures. The value of the product is reduced solely by the disfigurement and if the spots of mould are removed by careful brushing the fish may become saleable again. The best preventative measure is good housekeeping of the premises and surrounds by removing rotting vegetation and by having a regular cleaning regime. Dry, well ventilated and painted buildings should be employed if possible. Sorbic acid, where it is permitted by food regulations, is an effective preventative of this kind of mould; it is applied by dipping the fish briefly in an aqueous solution before stacking.

The storage life of salted fish depends upon the salt concentration, the moisture content and the prevailing atmospheric conditions. Other things being equal the lower the salt and the higher the moisture content the shorter the life. Light cures like the Gaspé last only a few days at ambient temperatures and must be kept near 0°C if a storage life of more than a week is aimed for. Hard dry cures will keep in good condition even at fairly high ambient temperatures for a year or more at low humidities. Storage deteriorations in salted white fish take the form of a gradual softening and develop-

Fig. 4.20 Replacement where possible of the traditional hessian package on the left by materials such as those shown on the right which are impervious to insect and other damage can considerably help to improve the quality of dried salt fish.

ment of off-odours due predominantly to enzyme action; if the moisture content rises to too high a level the fish will go bad as a result of bacterial spoilage.

Effective packaging can prevent to a considerable degree the spoilage consequent on the absorption of moisture from ambient air of high relative humidity. The best materials for this purpose are flexible plastic films, plastic containers or bitumen-lined brown paper. Salt fish is often sent to or prepared in countries with warm humid climates where insect infestation or rodent attack are prevalent and cause extensive losses. For these reasons, also, adequate protection is important, though providing it at a cost low enough to keep the product within the means of some customers is a severe problem. Extended storage life may be obtained by closely pressing the individual fish together in a mass to provide as near anaerobic conditions and as much protection against changes in humidity as possible. Pickle-cured partially eviscerated herring and similar fish remain in good condition for 6–9 months at average ambient temperatures, but gradually become unpleasantly rancid after that time, whilst corresponding whole herring keep rather less well, since the visceral enzymes (and possibly bacteria) cause extensive softening and putrid odours. It is not advisable to freeze pickled herring because extensive rancidity and discoloration can quickly occur.

(ii) Defects

Most of these follow from points already made: poor cutting, trimming or cleaning away of blood, viscera, belly-membrane; belly-burst; unattractive rough and salt encrusted surface and presence of bones (where these are undesirable quality features); discolorations; softening or excessive brittleness; wet patches; wrong salt or moisture content; bitterness due to presence of excessive calcium or magnesium; insect infestation. Further defects are: presence of non-permitted preservatives; excessive amounts of broken pieces of fish; off-odours and taints; faulty packaging including leaking barrels; mixing of different sizes of fish in the same consignment; contamination with sand, dirt; 'white spot' – disfigurement (caused by crystals of disodium hydrogen phosphate) avoided in the best quality dry salted white fish.

Measures to prevent these defects are self evident for the most part. If there is any doubt about the salt or moisture contents of the final product these should be checked. 'White spot' is associated with very dry conditions during salting and incipiently stale fish.

6 Dried fish

Here preservation depends on reducing the moisture content to a level where micro-organisms and most intrinsic enzymes become inactive. Thus the amount of drying has to be considerably greater than that required for salted products. Generally a moisture content of not greater than 15–20% is aimed at, this being the upper limit below which moulds will not grow.

The aim in production is to reduce the moisture content quickly enough to avoid the concomitant spoilage. By employing artificial means (drying tunnels, suspending over fires) this objective can usually be realised, but much of the world production is made by drying naturally under the influence of sun, wind or frost. In some cases (e.g. Scandinavian stockfish) climatic conditions are cool and dry enough to avoid much spoilage; in tropical or sub-tropical countries some spoilage is probably inevitable resulting in strong odours and flavours that may, in fact, be desirable features of the product. Control of quality changes during natural drying is obviously heavily dependent on a steady, reproducible climate and under small-scale, under-developed conditions is very difficult. Drying tunnels of various designs are available with which it is easily possible to produce dried fish of consistent quality. A final check on the end product can be made by accurate measurement of the moisture content as described in Chapter 6. The best method of retaining natural quality in dried fish is freeze drying but this very expensive procedure has so far only been applied to sophisticated articles like shrimps and prawns.

Raw material should be as fresh as possible, wholesome and if necessary gutted, headed, cleaned and washed. On the other hand, it may be necessary to avoid the use of split fish or fillets in the pre-rigor state because high temperatures of drying could induce vigorous contractions of the flesh with consequent tearing and a tough texture. If very fresh squid are dried too rapidly, black pigment cells in the surface layers of the body are liable to become prominent causing an unwanted dark appearance in the finished article. This can be overcome by soaking the dead animals in fresh water which causes the pigment cells to contract. For some products like Japanese dried bonito sticks ('fushi') the fat content of the raw material is quite critical and should be the moderate value of 1–2%. The ink sac and 'bone' in squid and cuttlefish are removed before drying.

Drying is expedited by creating as large a surface area as possible
– splitting large fish into thin sections; breaking or mincing;
spreading, hanging or laying out thinly; turning over laid-out fish
frequently. Under given conditions of temperature, humidity and
air flow, moisture escapes from the surface of the fish at an approximately fixed and steady rate. Making the surface area as great as
possible therefore maximises evaporation and minimises drying
time. Unless the product demands otherwise, initial removal of
some body juices by pressing out or by heating until the flesh
proteins coagulate is particularly advantageous. The water in raw
fish is held quite firmly and is accordingly removed very slowly.
Coagulation releases the water as a separate juice which then
evaporates more rapidly. Much use is made of these techniques in
the fisheries of the developing nations. Cooking before drying also
has the advantages that it normally avoids the occurrence of case-
hardening during rapid drying and arrests the actions of micro-
organisms and enzymes.

When drying naturally, protection against direct strong sunlight
is often necessary in order to avoid partial cooking and break-up of
the flesh or case-hardening; likewise cover from direct rainfall
which would wet the fish is an obvious necessity. If the surface
of drying squid becomes wet through exposure to rain or high
humidity it develops an undesirable reddish colour caused by the
leaching out of pigments lying just under the skin. Where humidity
or insects increase at night the fish should if possible be taken
indoors. Fish spread out-of-doors should be protected against
fouling by and depradations of birds and other animals.

Where means of measuring the moisture content are not available
it may be necessary to fall back on experience of how to produce a
stable product. Appearance and flexibility are rough guides to
moisture content.

(i) and (ii) Deteriorations and defects

Many of the points made under salt fish insofar as they relate to
dried fish are of direct relevance here and will not be repeated.
Properly prepared, dried fish should be free of more than the
odours and flavours characteristic of the product. Assuming that by
packaging or storage in reasonably dry conditions the moisture
content is kept below the figure already quoted, the product should
keep for several years at average ambient temperatures.

In humid conditions (that is when the relative humidity is greater than about 75%) unprotected dried fish of initially low moisture content will within a few weeks take up sufficient water to allow mould growth. A major problem in the control of quality wastage of dried fish in developing countries is insect attack. A fully economic solution has not yet been found in most cases. Good packaging where it can be afforded is one answer. Dipping before drying of fish in a dilute solution of the odourless insecticide pyrethrum, or light dusting of the finished product with the dried substance suitably diluted, both offer good protection against insect attack but should be resorted to only when other methods that avoid the use of insecticides have failed. Where it is feasible, fumigation of spaces where dried fish is stored is effective in keeping insect infestation under control.

The very slow deteriorations in non-fatty species are first the enzymatic production of off-flavours and secondly chemical reactions between carbonyl and amino compounds ('Maillard' reactions) that result in yellowish or brownish discolorations and burnt or singed off-flavours. With most dried fish products no attempt is made to control these deteriorations and they are only really important in the very small quantities of freeze-dried shrimps and prawns. Unless protected, dried fatty fish quickly becomes rancid and softens. Packing in an inert gas or vacuum offers good protection or, if they can be dispersed effectively through the product, antioxidants also considerably lengthen storage life.

7 Marinades

From time to time various groups of products incorporating acids, sugars or salt have been included under this title, but here the term is applied only to those preserved by the combined action of dilute acetic acid (usually in the form of vinegar) and salt. The inhibitory effects of these two substances on bacteria and enzymes are greater at higher concentrations but, since marinades are eaten without any further preparation, storage life is limited by the upper concentration, particularly of acid, that is palatable. Marinading is normally a two-stage process with the transfer of the fish in between. The aim in the first stage is to render the fish, normally filleted, rapidly as sterile as possible whilst at the same time developing the characteristic basic texture and flavour; for this purpose immersion for at least a week and frequently much longer in concentrated pickle solution is employed (5–10% acetic acid plus 10–15% salt). In the process the protein of the flesh is coagulated and the remaining small bones are softened. The aim in the second stage is to maintain a palatable level of preservative that will keep the product for a reasonable length of time. A final product in 1–2% acid plus 2–4% salt will keep in good condition for at least 3 months at near 0°C.

The eating quality of this product is very dependent upon freshness and lack of damage or contamination of the raw material. Frozen and thawed fish of good quality is quite acceptable. Fatty species and herring *par excellence* make the best traditional marinades; a minimum fat content of about 10% is required. Very fatty fish should be avoided if possible because an unpleasant looking and possibly rancid layer of fat may in time rise to the top of the second covering pickle. Salt-cured herring are also suitable after some of the impregnating salt has been removed by soaking in water. Marinades are a fairly costly delicatessen article and their appearance is of considerable importance. Thus, the fish should be cut, cleaned and washed in brine very carefully to remove traces of viscera, slime, blood and dirt. If boiled or fried fish are to be marinaded, this pre-treatment should be done thoroughly and in a consistent manner.

To ensure uniform and rapid penetration the whole fish or fillets should be mixed thoroughly with the first pickle, introducing them individually and stirring or agitating the mass from time to time. Fish in open vessels should always be kept below the surface, or closed containers used that can be kept completely full. Apart from effects on flavour and keeping quality, the texture of the final product is dependent on the concentrations of preservatives used; more acid softens, an effect that is moderated by increasing salt concentrations. Thus, the relative proportions of acid, salt and fish in both pickles is of great importance and should be strictly controlled for each type of product.

Fish that on removal from the first pickle are obviously spoiled or show discolorations cannot be improved by transfer to the second pickle and should be discarded. The quality of ingredients like spices and vegetables added to the final pickle requires special attention. If gelatin is used in the final jellied packing medium, it must be of food grade. Marinades are often packed without heat treatment in transparent jars or in cans suitable for retail display; all containers of this kind should be washed if necessary and inspected before use for damage or faulty closures. Fish should be trimmed and packed neatly to show to best advantage their silvery skin or added garnishes.

(i) and (ii) Deterioration and defects

If well-tried recipes and concentrations are adhered to there is no reason why a consistent, safe product of level quality should not be turned out. A useful safeguard against mistakes in processing that would affect quality is provided by occasional routine checks of acid and salt content in the end product as described in Chapter 6. Some bacteria and proteolytic enzymes are still weakly active in the second pickle and the product will gradually develop off-odours, discolorations and softening at a rate depending on the temperature of storage. Marinades cannot be frozen because break-up of the flesh can occur. Addition of small amounts of the bacteriocidal substance hexamethylenetetramine to the final pickle prolongs shelf life but is not permitted in many countries.

8 Heat processed fish

There is a considerable world output of various products that are boiled, fried or steamed but are not packed in hermetically sealed containers and it seems worthwhile to include some general remarks on them. The incidence of food poisoning is probably greater from this group of fish products than from any other.

Depending upon its intensity, heat processing will kill a varying proportion of bacteria and destroy most flesh enzymes. Thus, spoilage will result from the heat-resistant organisms that have survived the cooking procedure plus those that have subsequently contaminated the product; these latter occur during handling and packing. Nevertheless, the onset of spoilage will be delayed in comparison to corresponding fish that has not been heat processed. Unfortunately some food poisoning organisms are rather heat re-sistant and others are picked up by handling after cooking and may prove a hazard to health before the changes that result in obvious spoilage signal that fish is bad to eat. Therefore the precautions that must be taken to ensure with a high degree of certainty that heat processed fish is microbiologically safe are more stringent than those appropriate to raw fish. Cooking or pasteurisation should be carried out strictly according to procedures worked out and proved to be adequate, and more attention given to hygiene and cleaning both before and after heating.

Microbiological, including virological, testing is advisable to ensure that processing is adequate to render the product safe to eat. Re-testing is also advisable if the equipment or processing conditions used are changed. Because of the danger of cross-contaminating cooked warm fish with spoilage bacteria originating from uncooked fish, the processing of the two should be kept well separated from one another and preferably in different enclosed areas of the factory. As for other cooked foods, cooked fish should not be re-heated under conditions that encourage growth of food poisoning organisms. Cooked fish quickly loses moisture when hot and becomes unpleasantly dried out; it is bad practice to keep it for example in a hot cabinet for more than about 15 minutes before further preparation or serving. Cooked fish which is cooled slowly or kept warm for long periods can also develop a rancid or cardboard-like odour and flavour. It should be cooled rapidly by, for example, immersion in clean potable cold water or exposure to cold, moving air.

For some markets heating offers effective means of arresting spoilage. Shrimps are often cooked before further processing which allows the primary treatment to be carried out on fishing vessels. Cooking on board immediately after capture helps to preserve flavour and colour but because of the risk of subsequent bacterial contamination the product should be handled as quickly as possible. Sea water used to cool shrimps after cooking on board should be clean and uncontaminated. In some countries benzoic acid is permitted under regulations as a preservative for heat processed shrimps and prawns: clearly, the application of this substance and its final concentration in the product must comply with the appropriate regulations. Another method applicable to crab and shrimp meat is to heat them in suitably heat-resistant plastic pouches so that the centre is kept for 5–10 minutes at pasteurising temperatures (80–85°C). Most of the organisms are killed, and since the pouch offers good protection against re-infection a much enhanced storage life of 4–6 weeks at near 0°C is possible before the product becomes inedible. In the Far East a common preservative technique is to boil whole eviscerated fish in strong brine, to cool it and to distribute it without further handling. Storage lives of 2–10 days, depending on ambient temperature, are attainable and these can be extended by re-boiling after some days. The method could be improved by removing the cooked fish quickly from the preparation area to avoid cross-contamination and by more adequate protection from dust during distribution.

Cooking fish pre- or in rigor may present quality problems. As noted before, the flesh may break up and become very tough or rubbery. Also the flavour is often peculiarly metallic or watery and quite abnormal. Holding the fish for a few hours at chill temperatures until rigor has passed removes any difficulty of this kind.

Gelled products such as kamaboko, fish sausage and ham made by heat setting specially prepared mince are traditional and very important in Japan but as such are little eaten elsewhere. During the past decade the technology used in their manufacture has been transferred to the preparation of a variety of realistically simulated fish and shellfish products of which the most important is crab leg meat. These simulated products have found a ready sale in many countries outside Japan; manufacture of them has also been set up in several countries outside Japan. The details of their manufacture and of quality control are complex: specialist literature should be consulted if more information is required.

5   Further aspects of quality

1 Microbiology

Micro-organisms on fish and fish products can be divided into two groups depending upon their effect on quality: first, spoilage organisms the actions of which have already been described; and second, those of so-called public health significance (pathogens). The first group is, in practice, almost always present in fairly large numbers and are combated as part of normal quality control measures. The second group, although not often present in quantities that are hazardous, are clearly of more intrinsic importance and are, in fact, the main concern of microbiological quality control. With one exception (*Clostridium botulinum*), pathogens, in contrast to spoilage types, multiply only very slowly or not at all at 0°C–10°C, temperatures at which fish are commonly kept. Therefore, fish products present less of a hazard from this point of view than many other types of food.

Apart from the relatively very small quantity that is harvested in water contaminated with sewage, fish are in general free of intrinsic injurious organisms (the exceptions are dealt with later). Such organisms are, on the other hand, picked up during handling and processing, which is why cleanliness, hygiene and sanitation at all stages are so important. The following points which help to reduce the contamination are applicable to any food industry but are worth reiterating in relation to the fish sector: the proper construction of premises, equipment and drains; provision of adequate ventilation and for prompt removal of offal and refuse; regular, supervised cleaning and sanitation programmes; protection against rodents, insects and birds; supervised attention to personal hygiene and provision of pleasant, clean toilet facilities; provision of a regularly replenished supply of clean protective clothing and insistence on its use; proscription on smoking and spitting; attention to cuts and abrasions; separation of wet processing from dry packing areas.

Fig. 5.1 Clean, well-maintained toilet and hand washing facilities and rules governing hygiene are essential to the manufacture of high quality fish products.

The list is not exhaustive. Several of these points are covered by those provisions of local food hygiene regulations that apply to fish and fish products. It goes without saying that these should be complied with.

In many cases the efficiency of these measures in reducing the risk of contamination can be taken for granted, assuming they are applied with common sense and diligence. In a few other cases where the risks of mistakes or accidents must be reduced to a low level, for instance in a large factory or where critical processes (canning, pre-cooking) are carried out, it is necessary to monitor the cleaning and hygiene programme by regular bacteriological examination of equipment and perhaps even personnel. Also, the microbial content of the following may need regular checking: the water used for washing and processing; baths or sprays of additive solutions, brines etc.; batter used in enrobing. The growth of the microbial population should be prevented by keeping these media cool and frequently made up afresh.

A further more direct measure of microbiological control is to assess the degree of contamination of the finished product – and sometimes the fish and ingredients that are used as raw material. Since such assessments take many hours to complete, instantaneous or even very rapid action cannot be taken on the results. The value of the tests in terms of continuous checking of current end-product quality is therefore limited to products that are not distributed immediately but stored until the results are known. Even here the chances of consistently detecting hazardous samples are not high given a level of testing which is not prohibitively expensive. Microbiological control using cleaning, hygienic standards and sanitation during processing should be the first line of defence; end product testing alone should never be relied upon to provide surety.

Microbiological testing of end products is nevertheless of some value in that the results act as a warning that something is amiss in production; it can at least protect the safety of the succeeding batches of product. Foods already distributed but found on testing to be contaminated may be withdrawn from sale. In fact, most large frozen fish processors consider that it is essential to provide continuous, often daily, checking of production on the basis of samples selected according to a proper scheme. Action is taken when the degree of contamination rises above a specified level that, given a considerable margin of safety, is judged from experience likely to cause a food poisoning outbreak. Once an undesirable level of contamination has been found, detecting the avenue through which it arose requires an intelligent appraisal of where in the complex of operations this is most likely to be. When the detective work has been successful the avenue must be eliminated; raw material that is found to be unsafe must be discarded, of course. Microbiological testing for pathogens is an important means of protecting consumers and of ensuring that a buyer is receiving wholesome goods. Some countries or states acting through regulatory authorities insist that imported fish products should have a specified microbiological quality, though the specification is not always disclosed. Likewise, companies will lay down specifications for the guidance of suppliers. In both cases, samples of the consignment are tested either on despatch by the supplier or on receipt by the authority or company to make sure the goods comply with the requirement.

Since the group of organisms presenting a health risk grow best at 35°C–37°C, whilst spoilage organisms do not (their optimum temperature for growth is about 20°C), a measurement of the total numbers of organisms that grow when the product is incubated at this higher temperature gives a first approximation of the degree of contamination of the former. However, this figure has to be interpreted with caution and a comparison with the numbers that grow when the incubation is carried out at about 20°C is a safer guide. Almost invariably an examination for specific pathogens is also included.

As discussed in the section on intrinsic microbiological quality (Chapter 2), it is not practicable to determine all the pathogens that could be present in any given sample and one must resort to determining the incidence of certain indicator organisms, in this case *Escherichia coli* and coliform bacteria. However, on occasions it is useful as confirmation of a certain degree of contamination to measure the incidence of one or two of the more easily determined specific pathogens, namely *Staphylococcus aureus* or coagulase-positive *Staphylococci* and *Salmonella* species. It is beyond the bounds of possibility that all samples will have a zero incidence of all pathogens and a low level is unavoidable. Very little direct information is available on fish products that relates levels of incidence of pathogens with risks of outbreaks of illness. However, sufficient experience is available from foods generally to suggest that average counts of not greater than $10–10^2$ coliforms/g of product, 10 *E. coli*/g, or 10^2 *S. aureus*/g and an absence of

Salmonella in a given weight of sample (varying from 1 to 100 g) are safe working figures for fish and fish products. As a guide, figures of not greater than 10^4–10^6/g for the total number of organisms that grow at 35°C–37°C are suggested. Further recommendations on specific products are given in Appendix 5.

Should the numbers of organisms in product or raw material rise much above these figures, then action to correct the situation is justified. We are here touching on the subject of standards – a fuller discussion of how to view figures of this kind will be undertaken in Chapter 8.

As mentioned already there is no evidence that eating fish spoiled in the raw state is dangerous except in the case of a type of poisoning associated with species of the mackerel, tuna and saury groups (scombroid poisoning). Naturally occurring spoilage bacteria probably act on the plentiful amounts of histidine in these fish to produce biologically active amines that when ingested in sufficient quantities give rise to a rarely fatal reaction consisting of headache, dizziness, nausea, vomiting and urticarial eruptions. Often the incriminated fish is said to taste 'sharp or peppery'. The proper measure against causing illness of this kind is, obviously, to prevent spoiled fish of these groups reaching the consumer; normal vigilance on the part of the fish trades and official services should take care of this. Proper chilling and rapid distribution are the key elements. In Western Europe the incidence of this illness is very low indeed.

The two harmful organisms found naturally occurring in fish are *Clostridium botulinum* and *Vibrio parahaemolyticus*. The first occurs in soil and freshwater mud and may also be found in marine areas which are close to land and receive large volumes of fresh water draining from the land mass. It follows that freshwater fish are frequently found to be contaminated with a few *C. botulinum* organisms, and marine fish caught in the vicinity of large land masses may also be similarly contaminated. The organism causes no problem on raw wet fish since spoilage bacteria also present render it inedible before the pathogen can become dangerous. When it grows on food *C. botulinum* produces a powerful toxin that if consumed in even minute amounts affects the nervous system and often causes death. Fortunately the toxin is easily destroyed by heating – for example, in 2 minutes at 70°C in near neutral conditions – and even infected fish is safe if cooked. The danger arises with products that are eaten uncooked; there is a risk even with those that were once heat processed because they can either become reinfected or the spores of *C. botulinum*, which are quite heat-resistant, may survive and produce toxin. Fish that have been incriminated in the past include raw, canned, smoked and pickled products.

It should be emphasised that the disease is extremely rare and always associated with products that have not been processed according to well-established and proven techniques. An instance of the latter kind occurred when the salt content of smoked fish was progressively reduced over a period of time in response to public tastes for milder cures; the content was eventually reduced below the level that inhibited the bacteria. This illustrates the need for the microbiologist and those responsible for fish product quality to be alert to the consequences of changes in procedures that traditionally give safe products. In fact, the salt content in fish which is eaten without cooking should be at least 3% in the water phase in order to prevent production of toxin from *C. botulinum*. This guideline should be applied to products like hot smoked trout and mackerel, and cold smoked salmon. Normally the food will go bad before the amount of toxin is dangerous, but as pointed out previously this may not always occur. Theoretically, such conditions could exist if, for example, a fish product of low salt content was heat processed, packaged or irradiated then contaminated only with *C. botulinum* and stored for several days at above 10°C. For this reason processes such as vacuum packing, modified atmosphere packing and irradiation have been critised for introducing an enhanced risk from botulism. In practice, the enhanced risk is minute and is acceptable from the commercial point of view.

Additional safety is provided by freezing immediately after preparation and eating immediately after thawing or by keeping chilled below 3°C throughout, at which temperature the production of toxin is nil. The risk of botulism is so remote that it is not justified to test routinely for the presence of the organism. Only when the method of processing is radically changed, when fish suspected of harbouring the organism is used as raw material or when conditions exist that could conceivably introduce a much enhanced risk, is it advisable to conduct checks on the end product. In particular the capacity of the product to sustain the growth of the organism and to produce toxin under conditions that could be met with in practice should be examined.

Vibrio parahaemolyticus has been recorded chiefly in Japan where it is incriminated in 70% of food poisoning outbreaks caused by eating fish. In addition a few cases have been recorded in Australia, China and USA; it may be the unsuspected causative agent in other countries. It does not seem to inhabit cold sea areas. Generally the disease is mild. The organism is easily destroyed by heat, is susceptible to chill temperatures and is a mild halophile (it requires about 1% salt for good growth). Thus the risk of it surviving normal handling and preparation is very small; the predilection of the Japanese for fresh, raw seafood explains why they are particularly exposed to infection. In most countries there is no point in monitoring or checking for the presence of this organism; experts differ in their views as to whether there is a case for monitoring in the 'at risk' countries.

The hazards associated with pathogens in sewage and how they are dealt with have been described in Chapter 2.

2 Nutrition

In developed countries where adequate food is available for a properly balanced diet there has been up to the present time very little or no need to take special measures to ensure that handling, processing and storage do not impair the nutritive quality of the range of fish products considered here. In any case all the experimental evidence shows that the depletion in nutritive value accompanying normal commercial operations is very small or negligible. Fish is mainly eaten for its protein, fat, mineral and calorie content and these are virtually unaffected; severe or prolonged heat treatment as in canning or drying may deplete the concentrations of certain vitamins but this is accepted in the products anyway. Details of these losses are available in specialist texts.

An important aspect of nutrition as far as quality control is concerned is, however, labelling. There is a growing demand from consumers for more information about the nutritive value and composition of food to be shown on labels. Decisions about the nature and extent of that information are a matter of company policy, which in some cases will have to take into account legislative requirements. The role of those responsible for quality control is to help to ensure that the current information is printed on the labels.

This may require the fish or fish product to be analysed chemically and the results expressed in nutritional terms which it is hoped will be understandable by the consumer.

Scientific evidence is accumulating which suggests that eating a diet relatively rich in certain of the lipids (the chemical term for fats) present in fish can have beneficial effects in protecting against heart disease and possibly other diseases. Although very important for the marketing of fish, this evidence has no direct implications for the control of quality.

The developing countries are very concerned to get the utmost out of what food is available and with them there is more incentive to reduce nutritive loss to a minimum. The main way in which this can be achieved is, of course, by eliminating gross waste wherever it occurs – utilising every portion of the catch, maximising yields, eliminating spoilage and depradations of animals and insects, avoiding excessively severe treatments (for example, overheating during drying over open fires or in the sun). Waste of fish is very widespread in developing countries, most of it not wilful or deliberate but because of lack of knowledge, skills, money, equipment and social infrastructure. Those responsible for the control of fish quality in these situations have a critical role in reducing waste and so improving nutritional status.

3 Additives, preservatives, flavouring agents and colouring matters

The addition of chemicals or other adjuncts to improve the keeping quality or consumer appeal of foods in general and fish products in particular is accepted practice. However, the last few decades have seen vigorous efforts in many countries to limit, control and codify the use of all such substances. Most countries now have a corpus of food laws regulating these matters. It is the responsibility of those concerned with fish product quality to be alert to the requirements of the countries in or with which they trade. A wise trader will not risk loss of goods and goodwill or possible prosecution by allowing non-permitted substances in his wares. Efforts are also being made internationally with some success to harmonise food laws partly with the aim of facilitating trade. With the objective of reducing unexpected risks to health there is a general tendency to reduce the number of substances added to food. The situation is constantly changing and if there is any doubt about what is permitted it is advisable to consult the relevant literature (examples are given later) or specialist authority.

The traditional preservatives including salt, vinegar and acetic acid, alcohol and natural smoke are universally acceptable, though even the last-named has come under scrutiny on account of possible carcinogenic effects of some of its trace constituents. Very few 'artificial' preservatives (e.g. sulphur dioxide, sorbic and benzoic acids, hexamethylenetetramine) are now permitted in fish products; many countries do not permit any, and the possibility of new ones being generally accepted is, as noted previously, highly unlikely. As far as permitted flavouring and colouring agents are concerned the requirements for fish are almost always the same as for food generally, though one list (EEC) permits Brown FK only for kippers and no other food. With some major national exceptions polyphosphates are accepted for fish and fish products. In all cases the maximum amounts of added substances may be specified.

4 Yields, weights and microelectronic control

The growing cost of fish and the need to recover as much valuable protein as possible encourage greater attention to improving yield of edible portion. This task is the immediate responsibility of the production manager or similar employee but those responsible for quality also have a role to play in that yield is very dependent upon the correct selection of raw material and how it is treated up to the point of processing. Two intrinsic factors that influence yield are the size of fish (large sizes give a higher yield than small, other things being equal, processing machinery gives maximum yield only with certain size grades) and the biological condition (plump fish give a higher yield than thin). In terms of deteriorations and defects: spoiled or damaged batches will clearly give lower quantities of usable fish; well chilled fish often give higher yields on machine processing than warm; weight losses can occur as a result of crushing during iced storage or through dehydration in cold storage or through 'drip'. Corrective action for all of these has already been explained.

One characteristic of high quality products is the consistency with which units in a consignment meet the declared weight or quantity. This is achieved by careful control of fill or packed-out weight, making allowance for losses likely to occur during distribution and storage by adding the correctly estimated overweight. Production can be checked by regular, random sampling before packing followed by weighing or visual inspection. It is normally obvious where in the process errors of fill or weight are occurring and require correction. Machine filling or weighing can be complemented by regular inspection of the recorded weights or by automatic rejection of faulty units. Customers are quick to detect underweight or wide fluctuations in contents. Also, the food regulatory authorities in many countries keep a close watch on incidence of underweight in retail packs. On the other side of the coin the persistent giving of unnecessary overweight obviously should be guarded against. For all these reasons it pays to keep a close watch on accuracy and uniformity in the weight or count of goods.

Computerised logging and control of weights, counts, quality factors, yields and prices is an increasing feature of larger factories and, with the ever decreasing cost of computers, could even extend to small processing units. In addition, there is a growing range of other kinds of microelectronic devices that can facilitate the control of quality and processing operations. Quality control staff should be aware of these developments and should take steps to introduce any that would assist their work.

5 Packaging and labelling

Most food products sell on visual appeal and since the packaging is often all that is visible, its quality must be consistently high. The subjects of packaging design and the selection of the correct packing material for particular uses are outside the scope of this book and are mentioned only to draw attention to the fact that they are as important to fish products as they are to any other food. Several points on packaging have been mentioned already: contamination with undesirable substances from packaging materials; suitable moisture and gas permeability; correct assembly, sealing and seaming. Since consumers are aware that packaging is meant to protect the product they will reject or mark down damaged containers and goods with torn or misapplied labels. Removal of goods having faults of this kind before they cause complaints can be assured through a proper programme of visual inspection of the end product and of products on display.

Labels should correspond with quantity, quality and kind of contents. They should provide enough information to allow the buyer to understand the nature of what he is buying, and should not misinform. In some countries the way in which food is described is covered by regulations: even the size and arrangement of lettering and the exact words to describe fish species may be specified.

Methods of assessing and selecting for quality

In the preceding chapters various aspects of quality have been described. In order to control and select for these aspects it is necessary to have means both of recognising quality and of measuring the environmental factors that influence quality. An ability or willingness to control implies that there is some measure or standard in the background against which it is possible to take action. Thus to say that fish is of poor freshness quality and rejectable on that account implies that one has a firm measure of freshness quality against which the rejected fish is being matched.

In this chapter methods of measuring quality attributes and relevant environmental factors are discussed. A great many methods have been proposed or tested for measuring fish quality in industry. Many have been shown to have shortcomings like inaccuracy or impracticability serious enough for them to be rejected. Others are suitable only for research or product development. Only those methods will be included here that are practical and are either known to be in current use or in the opinion of the author are on the point of being introduced. Full details of methodologies are not given here; they may be found in the further reading list. It is convenient to divide methods into three groups having reasonably clear boundaries, and to then consider relevant statistics.

1 Sensory methods

These are defined as those wholly dependent upon the human senses perhaps aided occasionally by simple devices like a ruler. All the senses with the exception of hearing are used throughout the fish industries to judge quality – sight, touch, odour and flavour. The consumer obviously uses only his senses in deciding what he likes. It can be argued therefore that where they can be used sensory as opposed to non-sensory methods offer the best opportunity of getting a valid idea of what the consumer wants. Sensory methods also have the great advantage that human beings are very adaptable and can switch easily from, for example, testing odours to visual inspection for defects. Furthermore, for some tasks human senses are better at recognising complexities and are more discriminatory than instruments. Their main disadvantages are that their responses can vary, particularly with fatigue or outside distraction, and that using people can be expensive and inconvenient. Frequently there is no choice but to use sensory methods but some instruments like thermometers are also indispensible; microbiological methods stand on their own. Where either sensory or instrumental methods can be used to make the same kind of measurement, the choice of which to use will depend on the balance of advantage.

(i) Quality factors open to sensory methods

The uses to which sensory methods can be put may be grouped according to the sense being used.

(a) SIGHT AND TOUCH

The selection or sorting for species and size, basic to any fishery, fall under this heading. It requires a minimum of training to get fishermen to segregate by appearance the catch into species that are worth keeping and into the size groups that are wanted. After a little practice it is surprising how rapidly and consistently operatives can sort fish of reasonable size. In some port markets size grading of certain species is carried out by placing the fish individually against a marked board. Probably sensing of weight is brought into play when sorting by hand for size. Hand packing of cans or bottles requires an ability to select fish of a suitable size and weight to fill the container within a close approximation of the correct total weight of fish. Cutting fillets into pieces of specified weight is necessary for some trade purposes and with some experience can be achieved manually with a surprising degree of accuracy and precision. A specialised application of the same kind is the identification of species of canned salmon and other fish from the visual appearance under a lens of the scales.

All the visible signs of deterioration (for example, loss of freshness, cold storage changes) noted earlier are detectable by untrained or trained people, and it is unnecessary to repeat them here. In almost all cases the detection of deteriorations and defects is accomplished very efficiently and rapidly by sight. Instruments or machines have little or no place as yet in this area. For example, it would be very difficult if not impossible to make a machine capable of detecting poorly trimmed fillets, or fillets possessing too large an area of skin. The detection of parasites and pin and other bones is still largely carried out by sight or by feeling with fingers at those places where bone fragments are known to occur. Detection can be aided by illuminating the fish or cut surface with ultra-violet light: parasites and bones fluoresce more brightly than the surrounding flesh. In recently developed equipment the bright images can be recorded on a video camera and measured or enumerated automatically. For the assessment of degree of gloss on articles like smoked fish the practised eye has so far proved better than any instrument. A reasonable assessment of oiliness in fatty fish is possible by inspection of the raw fish and by tasting it cooked but here instrumental measurement where it can be carried out offers a surer answer.

In assessing textural attributes (firmness, softness, mushiness, rubberiness, woodiness, mealiness, succulence, dryness) the sense of touch in fingers or mouth are both used as occasion demands. If a tasting test on cooked or prepared fish is being conducted it is sometimes convenient to include in it an assessment of texture as well as of odour and flavour. For the most part there is no substitute for sensory methods in the assessment of fish texture though, as noted later, some instruments are available for measuring degree of firmness. The matching of colour can be done very effectively by eye. Grading of salmon, tuna, laminated or mince blocks and fried breaded products is aided by comparing their colour with a set of tiles or cards carrying different permanent shades. The use of instruments to measure the colour of fish and fish products is limited because of the technical problems involved.

(b) ODOUR AND FLAVOUR

If we confine ourselves strictly to responses in the mouth, the sense of taste is limited to a few basic notes. In everyday use, however, flavour is usually meant to include much of what is experienced on smelling through the nose. Thus, as far as the products under consideration are concerned odour and flavour can be taken together. These senses are powerful tools in assessing quality.

Anyone can, without hesitation, distinguish between the smell of fresh and bad fish. With some practice the whole pattern of changes in odour between very fresh and very spoiled can be differentiated easily and rapidly so enabling, as we shall see in more detail shortly, the degree of freshness to be accurately determined. Similarly, off-odours (cold storage, rancidity), taints and unusual intrinsic odours are readily detected and their intensity judged reproducibly. Up to the present time there is no substitute for the human nose as detector and assessor in any of these situations.

Assessment of degree of smoke flavour or smoke character is sometimes called for. Although the quantitative measurement of certain chemical constituents derived from smoke can be used in some circumstances to measure smokiness, tasting is the most reproducible method taking all types of product into account and is capable of assessing far more satisfactorily the changes in character of the flavour.

The four basic tastes detectable by the tongue – salt, acid, bitter and sweet – are in a somewhat different category. The separate assessment of saltiness and acidity are important in a number of products. Different levels of these tastes can be distinguished easily so long as comparisons are possible but the measurement with accuracy of degree of saltiness or acidity in an absolute sense is difficult. That is to say, it is easy to distinguish between 1% and 1.5% salt by comparison testing at the same time but it is difficult to identify in isolation and without a comparison standard whether a product contains 1% salt. Furthermore, the reaction to saltiness depends to some degree on how much fat is present. Here the chemical determination of salt or acid provides a more quantitative and precise measure. The agents causing bitterness, as in stale fish, have not been fully identified and thus cannot be analysed chemically; in this case the sense of taste has to be relied upon. The need to assess degree of sweetness is infrequently required but again tasting is the choice.

(ii) Types of sensory assessment

Depending upon how they are used, two types of assessment can be distinguished. The first is the dispassionate, unbiased, descriptive assessment of individual or groups of quality factors.

Examples of responses in this kind of assessment are: this fish is fresh . . . is salty . . . is stale . . . is sour . . . is tough . . . is sandy . . . has too many bones . . . is too pale . . . is the right size . . . breaks up on skinning . . . will not suit my customers . . . is up to standard.

Sometimes this kind of sensory method is called objective because the person carrying it out tries to remove from his judgement all feelings of liking or disliking.

The second is where, after examining the product, feelings of liking, pleasure, acceptance, valuation or prejudice are allowed full rein, as for example: I think this fish is excellent . . . is inedible . . . is OK . . . is unacceptable . . . I would buy this fish.

Judgements of this kind are termed on occasions either subjective because they are entirely personal, or hedonic because they relate to pleasure or degrees of it. The utility of subjective sensory methods in practical quality control is very limited because the result does not help much in deciding what corrective action to take if something is wrong. Being subjective the result is essentially fickle and its interpretation problematical. In quality control we are trying to get a measure of a particular aspect of quality which can be compared to a standard. A personal expression of like or dislike is not a useful measure of this kind. On the other hand, hedonic judgements can be used to define initially the quality of products that consumers like and which in production and marketing should be aimed at. This is done by getting groups of fifty or more ordinary consumers to express free-ranging opinions on products.

The descriptive, objective method in which error is reduced to a minimum is of much more use. If necessary the attributes of quality that consumers want can be related to the results. The latter can be standardised or matched against existing standards to provide useful information on which to take corrective action.

(iii) Scales, scores and grades

For many purposes in the selection for and control of quality it is sufficient for the sensory assessment to remain very simple – for instance, visual detection of occasional samples having an obvious defect in the majority of samples having none.

However, sometimes we are faced with making selections from fish having gradations of quality, defects or deterioration. Thus it is often necessary to sort by hand and eye a mixed lot into different size categories. In this case the sorter must have in his mind a clear idea of the boundaries of the categories. Similarly brining will often result in products with varying salt contents; tasting will reveal those that are too salty for the particular market. Here the taster must be able to appreciate different saltinesses and know what the excessive level is. In the case of spoilage at chill temperatures appearance, odour, flavour and texture pass through various well-defined stages which an inspector must be able to recognise, perhaps in order to condemn fish that has passed beyond a certain stage.

Other deteriorative changes also occur smoothly and continuously through varying degrees of intensity. Thus cold storage firmness and off-flavour progressively increase from being slight through moderate to extreme. The taster must assess these different levels of intensity and may act by rejecting those fish having a degree of deterioration above a certain level.

Wherever a graded series of changes of this kind has to be contended with it is useful to the person carrying out the assessment to construct a scale showing exactly how the changes occur. A scale provides the assessor with a fixed yardstick which he can use on different occasions. Others can learn or become familiar with the same scale and thus a common measure is established. Scales consist of a series of steps, each being described in a distinct form of words. For example, a scale of saltiness would run from 'no salt' through 'moderate salt' to 'very salty'. About four or five steps would normally suffice in a scale of saltiness, or for that matter most other attributes or defects. It is always possible to construct a scale showing change or incidence by the use of value words: slight, trace, medium, moderate, very highly.

A well known scale (somewhat abbreviated) showing stages by which the odour changes in spoiling white fish is the following; the description at the top relates to absolutely fresh fish, the others to decrease in freshness to absolutely putrid at the bottom:

Fresh seaweedy
Loss of fresh seaweediness, shellfishy
No odours, neutral
Slight musty, mousey, milky, caprylic
Bready, malty, beery, yeasty
Lactic acid, sour milky, oily
Acetic or butyric acid, grassy, slightly sweet, fruity
Stale cabbagy, turnipy, wet matches, phosphene-like

Amine, 'byre-like' (ortho-toluidine)
Hydrogen sulphide, strongly ammoniacal
Indole, faecal, nauseating, putrid

In this case the steps are for the most part not gradations but quite different in character.

A further development of scales is to denote the different steps by numbered 'scores' from 0 or 1 upwards; the spoilage example just given can be made into an 11-point scoring scale. This is not only a convenient shorthand device for identifying the steps but allows the results from different sensory assessments to be pooled, averaged and so on by the processes of ordinary arithmetic and statistics. If the numerical sensory spoilage 'scores' of say six fish taken representatively from a batch are averaged, the result can be said to describe the freshness quality of the batch. Depending upon where the acceptable quality level is set the batch may then be accepted or rejected according to where its average freshness falls. Sometimes it is possible to measure with a ruler, estimate by eye or count the degree of incidence of defects in a fixed amount of product. For example, the area of black belly membrane on a fillet can be measured or estimated, the number of fins counted. Such values can also be arranged in a scale the points on which may be 'scored', in numbers as before.

Since defects and degrees of deterioration are undesirable it is usual to think of their increase as warranting the award of demerit points or scores. Thus for a kilogram of skinless fillets the demerit points for 0, 3 and 6 cm^2 of skin might be 0, 1 and 2. Lots of fillets with the highest numbers of demerit points would obviously be worst. There are a number of essentially similar ways of handling scores of this kind and examples can be found in the standards and specifications reproduced in the Appendices. Where a number of defects or deteriorations occur together in the same product, it is common practice to sum the scores or points allocated to the individual attributes to give an overall demerit score or grade (a term explained below). Some defects are more objectionable than others – for example bones are worse than blood clots in fillets supposed to be boneless – and more demerit points would be awarded them in the overall assessment.

Because the intrinsic quality of fish is so variable and because workmanship is never consistently perfect, it is necessary when setting the scoring or points level above or below which the product is unacceptable, to allow a tolerance; a series of fish products without defects is an impossibility.

Grades have essentially the same meaning as scores but in most usage tend to be simpler, less finely subdivided ('disaggregated') or amalgamations of several quality attributes considered together. A Grade 1 or Grade A product would possess few defects or signs of deterioration of any kind; a lower grade, several of one or more kinds. Indeed, grades are often defined in terms of the total number of defects or demerit points. An example of a freshness grading scheme is given in Table 6.1.

Table 6.1 A freshness grading scheme used in the European Economic Community for whole, chilled cod, haddock, whiting (*Merlangius merlangus*), ling, hake, saithe and redfish (*Sebastes* spp.) based partly on the freshness odour scale given above.

Grade	Extra	A	B	C (unfit)
Skin	Bright, shining, irridescent (not redfish) or opalescent, no bleaching	Waxy, slight loss of bloom, very slight bleaching	Dull, some bleaching	Dull, gritty, marked bleaching and shrinkage
Outer slime	Transparent or water white	Milky	Yellowish-grey some clotting	Yellow-brown, very clotted and thick
Eyes	Convex black pupil, translucent cornea	Plane, slightly, opaque pupil, slightly opalescent cornea	Slightly concave, grey pupil, opaque cornea	Completely sunken, grey pupil, opaque discoloured cornea
Gills	Bright red mucus, transluscent	Pink, mucus slightly opaque	Grey, bleached, mucus opaque and thick	Brown, bleached, mucus yellowish grey and clotted
Peritoneum	Glossy, brilliant, difficult to tear from flesh	Slightly dull, difficult to tear from flesh	Gritty, fairly easy to tear from flesh	Gritty, easily torn from flesh
Gill and internal odours	Fresh, strong seaweedy, shellfishy	No odour, neutral odour, trace of musty, mousy, etc.	Definite musty mousy etc., bready, malty etc.	Acetic, fruity amines, sulphide, faecal

Fig. 6.1 An important stage in the control of quality is the official inspection that occurs at port markets.

In order to be classified as Extra Grade, for example, a batch of fish needs to possess all the characteristics in the first column of descriptions.

There are several examples of the same type for this and other products. Such grading schemes are useful whenever the quality of batches of fish has to be assessed rapidly, such as at port markets or factory reception areas. The inspector scans the fish taking into account more or less simultaneously a number of different attributes before allocating a grade.

(iv) Judges

In the sense used here a judge is anyone called upon to carry out sensory assessments. As noted before the use that can be made in quality control of ordinary consumers as judges giving hedonic opinions is very limited. Hence most judges must be experienced or trained to assess quality objectively. Much entirely adequate control of quality is done in all fish industries by a single expert judge. He might be the long-established owner of a small pro- cessing business or shop, the fish buyer of a large company, or a fisherman; he may not have had any formal training. His long experience of the requirements of his customers will often equip him with all that is needed to select correct and consistent quality or to spot inadequacies or mistakes in handling, processing or distribution. Increasingly, however, with the growing complexities of operations and greater technical demands self-taught trade experts are being replaced by individuals specially trained objectively and technically for certain quality control or inspection functions. The single trade expert is known to suffer from other disadvantages; his judgement may vary over time depending on the state of trade and other influences, any prejudices he has clearly have a major effect on the nature of the decisions about quality, and he may be difficult to replace.

Some of these disadvantages can be overcome at least partially by training the single judge in a systematic inspection, scoring or grading scheme. His judgements are then standardised, probably surer and can be checked if necessary from time to time. It is likely that the strategic introduction of trained staff given or with access to the necessary authority to effect change represents one of the best and cheapest ways of improving sensory assessment and quality. The word 'strategic' is important: as will be made clear this does not imply the recruitment in any single enterprise of large numbers of persons trained in sensory methods who then continuously monitor production.

Further improvements in sensory methodology are possible by having two or more trained judges, normally acting independently but assessing the same product. By this means the risk of major mistakes through prejudice or bias is almost entirely eliminated. If all the judges assess the same sample or different samples from a batch, using the same points scoring system, the averaged results give a surer measure of the attribute in question than is possible with a single judge. The maximum number of judges required in this type of work is about six. A group of judges is often called a taste panel, particularly when their judgements involve tasting. Whether a taste panel is justified depends upon how much is at stake as a consequence of letting samples through of a quality that is below the required standard. A large company making a branded fish product on a considerable scale may have a very large investment at risk and consequently would be justified in employing a taste panel to conduct checks on raw material or product – always assuming, of course, that it could be afforded. Clearly, the use of

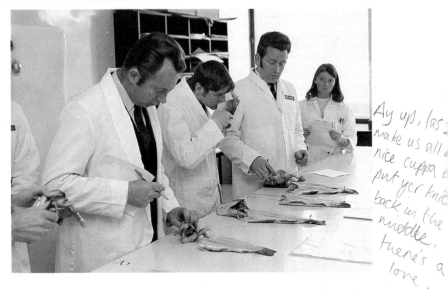

Fig. 6.2–6.3 Panels of judges assessing in the laboratory the quality of fish and fish products by means of appearance, odour and flavour.

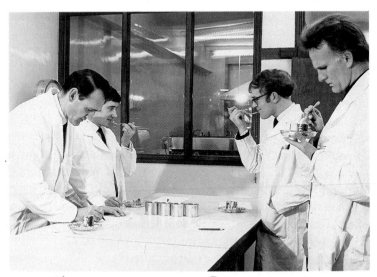

trained personnel and taste panels should be confined to the subtler, more difficult to measure attributes; it would be pointless to employ such talent on, say, the simple evaluation of defects.

(v) Precautions — Discharesy

Sensory judgements are sometimes influenced by external factors. The colour of a product will appear different under different types of lighting. Therefore in order to carry out careful colour comparisons as standardised lighting conditions as possible should be used. The detection of a special odour in a working environment like a market can be hampered by extraneous odours; practice is therefore usually needed so that the unwanted interference can be ignored. If samples are tasted after heating or cooking, the process should be carried out in an identical fashion on every occasion: it is no use trying to compare the level of off-flavour in fried with that in boiled fish. Similar considerations apply to test canning carried out to determine the suitability of batches of raw material. Sensory assessments to be at all precise require a certain amount of concentration, so distractions should be avoided. Ideally tests should be conducted in an area or separate room isolated from processing and other industrial operations.

2 Mechanical, instrumental and laboratory methods

These three methods have the advantage of not being subject to human vagaries and therefore are more easily reproducible and reliable. Although there is some overlap between them it is useful to distinguish two types of method: mechanical/physical and chemical/biochemical.

(i) Mechanical and physical methods

The problem of sorting or selecting for intrinsic quality factors is in some instances open to physical solutions. First, machines for size grading are available, but not yet unfortunately for sorting different species or for sorting food fishes from trash. These machines can sort many round and certain types of flat fish species and shrimps and prawns at a rate and finely enough for most commercial purposes. Equipment is available that will weigh fillets or pieces of fish automatically and either sort them into different weight ranges or assemble them into batches of the same weight. Second, two machines based on different principles are available for separating males and females of herring and capelin. Whether to choose machines in preference to hand sorting depends on factors such as the availability of space and electricity, of capital and running funds, of labour, or maintenance facilities. Third, a number of instruments for measuring moisture and fat are available. Determinations of these parameters are done most accurately chemically, but automatic or semi-automatic instruments are available that do the job rapidly and to an accuracy sufficient for many industrial purposes. A device for moisture consists of an infra-red heater to dry quickly the finely comminuted sample which is weighed simultaneously. For fat determination the minced sample is rapidly extracted with a fixed quantity of a fat solvent (tetrachloroethylene) by intensively shaking them together, the solvent is drained off and its specific gravity measured automatically with a special balance. The specific gravity is proportional to the amount of fat in the solvent. Protein and fat content in some fish products can be meas-

Fig. 6.4 A British instrument for measuring conveniently and instantaneously the degree of spoilage in chilled fish; the meter reading appears as an illuminated number in the window at the upper part of the instrument.

ured rapidly by an instrument using near infra-red radiation, but it suffers from the drawback of requiring calibration for different products. Furthermore, its expense would rule it out for any but major companies. Fourth, expensive instruments are available for measuring various aspects of colour: again their application is limited.

When we come to deteriorations and defects in raw material and products, a few physical methods are available. Progressive and marked changes in the electrical properties of the skin and underlying tissue provide a means of measuring the degree of spoilage in most types of chilled whole fish. Two different direct reading, instantly-responding instruments based on this principle have been developed. The older is German, the newer British. They provide a meter reading that can be calibrated in terms of the degree of

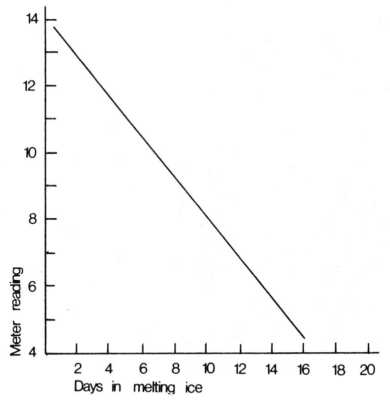

Fig. 6.5 The change of reading of the instrument shown in Figure 6.4 with the degree of spoilage of cod; the exact form of the change differs somewhat for different species.

spoilage or the shelf-life remaining before the fish is inedible. Figure 6.5 shows the typical change of meter reading of the British instrument with period of storage of cod in melting ice. It is small enough to be carried around in the hand for direct application to the surface of the fish. No damage is created to the sample by the test. Satisfactory readings cannot be obtained with frozen and thawed fish, fillets or flesh. The main areas of application of such instruments are therefore at port markets and factory reception areas where they can aid the buyer, factory manager or quality controller in rapidly checking on the freshness quality of batches. The first design of instrument suffers from some disadvantages not present in the second and perhaps for this reason the former has not been widely taken up by fish industries. Nevertheless, a reliable, robust and accurate instrument of this general type could provide a very useful standard tool for checking and monitoring, and it is believed that the British design fulfills these requirements. Its advantages are that it is independent of fallible human judgement, does not require highly trained or experienced personnel, never tires and in some circumstances can test more fish in the same time than is possible by sensory methods. Its main disadvantage is that readings on single fish are by no means a sure guide to freshness as judged sensorily and for this reason it is essential to test several fish in order to obtain a good measure of the average freshness of a batch. Equipment based on the same principles has been constructed to measure automatically the freshness of individual fish moving along a conveyor belt, but this is not in widespread use.

A number of instruments for measuring firmness in frozen, chilled and canned fish have been designed but so far little use for them has developed. The objective is to have a reliable, standard device for detecting samples that lie outside the range of normal acceptability. If this can be done, action to correct the process or selection of raw material can be taken on a rational basis. Two basic types of firmness instruments are those that measure the force necessary to press a plunger or set of sharp needles into the sample, and one that measures the degree of fragmentation when fish flesh is subjected to a fixed amount of disintegration in a homogenising apparatus. The former are in use in some salmon canneries. In the latter, designed for frozen products but not yet in commercial use, the degree of fragmentation (measured by an optical method) is less in firmer specimens and thus provides a measure of textural eating quality. In all these instruments the sample is destroyed during testing. Because of considerable variability several samples must be tested to provide an average value for a batch; single readings may give a misleading impression of the firmness of a sample. As pointed out earlier the pH of fish flesh is directly related to firmness. The direct measurement of pH on raw material before acceptance for processing has been proposed as a means of avoiding fish that could be too tough or otherwise of undesirable pH. So far little use outside the canning of albacore has been made of instrumental determinations of pH though technically it is possible. The measurement requires a reading of the electrical potential between a so-called glass electrode and a calomel electrode, both directly inserted into the flesh.

Determinations of drained weight, net content or 'drip' require only simple apparatus (standard sieve, balance, clock) but because of their relatively empirical nature should be carried out carefully under standard conditions keeping such factors as temperature and drainage time constant. A number of methods suitable for different products are available; the method used should always be noted in reports.

Sometimes it is necessary to check the fish content of battered or breaded products like fish fingers. The simplest way of carrying this out is to carefully remove the batter or crumb coating after softening it in water. A direct comparison of the weight of the fish centre with the total weight of the finger provides a measure of fish content. Stripping off the coating is a somewhat subjective procedure. A precise but more laborious method involves the measurement of protein content as described in the next section.

Various physical methods are available for the measurement of salt content, but in every case the salt must first be extracted from a given weight of the product by homogenising it in a measured amount of water. The salt concentration in the filtered or centrifuged water can then be measured by a conductivity meter, an ion-specific electrode instrument or, rather more inaccurately, by dipping in a proprietary strip of paper (Quantabs) whose colour changes depending upon salt concentration.

As mentioned above, parasites and especially bones in fillets and thin pieces of fish can be located, measured and enumerated by a detection instrument using ultra-violet fluorescence. This equipment is capable of grading fillets or pieces according to the number and size of bones detected and may be eventually capable of guiding devices for cutting out unwanted defects of this kind.

Metal detectors are in common use for screening products such as fish fingers and portions just before packing.

Small differences in the refractive indices of the oils from different species allow a tentative identification of an unknown sample of fish to be made. For this the oil is extracted from the sample, for example canned salmon, and its index determined in a refractometer. The technique does not allow an unequivocal identification and additional information is necessary, for example on the appearance of the scales.

For reasons that have been stated, the control of processing time and temperature is of vital importance in ensuring the maintenance of quality. It can be claimed justifiably therefore that the clock and thermometer are the two most important physical instruments. Their careful and sensible use can go far to obviate the need for other more advanced instruments or expensive methods of inspecting and checking.

(ii) Chemical and biochemical methods

These are applied in two areas which will be dealt with separately (a) composition of raw material and products; (b) deterioration.

(a) COMPOSITION

Occasions are small in number where determinations of water content (more generally moisture content) are called for in the range of products considered here. They may be necessary for example where the presence of excessive water in fish raw material is suspected or where it is important to dry a product to a specified water content. In these circumstances a representative sample is weighed out accurately and dried in an air or vacuum oven at a temperature sufficient to drive off the water in a particular time (conveniently 5–20 hours). After a prescribed period the sample is re-weighed. The procedure is somewhat arbitrary in that some damage and consequent weight change of the dry matter in the sample is inevitable. However, this complication is kept to a minimum by the adoption of suitable mild, standard conditions of time and temperature. These conditions should be stated when reporting the results.

Likewise, determinations of protein content are seldom required.

One instance is in connection with the measurement of the fish content of a product. A minimum fish content of some products like cakes and fingers is specified in the food regulations of some countries. The method depends upon knowing the protein content of the fish used in the product and if necessary making a correction for the protein contained in other ingredients. By simple proportion the amount of fish in the sample can then be calculated. In practice the method is subject to a number of uncertainties and is only capable of detecting samples showing rather wide departures in fish content from a specification. Protein content is assessed by determining the nitrogen content of the sample and multiplying it by a factor (generally 6.25) representing the inverse of the known nitrogen content of the protein. Nitrogen is determined by the Kjeldahl method.

As stated above, chemical methods for fat are preferred where it is necessary to obtain the highest accuracy or most general application. Here the weighed sample is extracted with a fat solvent in a special refluxing apparatus and the solvent evaporated from the resulting solution of fat leaving a dry residue of the latter which is then weighed.

There is an occasional requirement for measurement of the total mineral content because this is a good measure of bone, shell or sand content of a product. Bone or shell content may need to be controlled in fresh or edible material recovered mechanically from trimmings, skeletons, claws and the like. The mineral content is determined by burning off at a high temperature the organic part of a known quantity of the product and weighing the residue of ash.

Since salt (sodium choride) is an important constituent of many fish products, determinations of its concentration are frequently necessary. Chemical determination of salt is more accurate than the physical methods just described. It is carried out by dissolving the whole sample in nitric acid and measuring the concentration of chloride in the solution by titration with silver nitrate solution.

In acid-preserved fish like marinades it is necessary during production to check from time to time that the product is sufficiently acid. This is achieved by titrating a sample with a standard solution of alkali after comminution in a known volume of water.

The relevance to quality or safety of several other constituents has been mentioned but in most cases determinations of their concentrations will not concern the industrial control laboratory. These constituents include metals (mercury, lead, cadmium, zinc, tin), chlorinated hydrocarbons, radioactive isotopes, colouring matters, additives and preservatives. Analytical methods for these are complex and are normally only undertaken by specialist laboratories. If firms have need to determine constituents of this kind they would be advised to consider arranging for the analyses to be done by a consultant. Government or local authority laboratories need to be able to determine some of these constituents in connection with their normal food monitoring duties.

A further special compositional method that is perhaps of more widespread importance relates to the identity of fish species in samples. A growing number of-standards and regulations demand that the species of fish in products be named correctly often according to unique, designated names. The objective is the avoidance of fraud by substituting inferior species for more highly valued. An example drawn from UK legislation is the Labelling of Foods Regulation (1970), Statutory Instrument No. 400 (and amendments). Also, purchasing firms wish to ensure that fish not otherwise identifiable – say, skinless fillets – is of the species ordered. In these cases the recommended method is to extract the proteins of the fish into an aqueous solution and then to separate them in a process known as gel-electrophoresis which provides a pattern of bands of different intensities. Every species of fish has a uniquely different pattern that serves unequivocally to characterise it. In this method, which exists in several variants, unknown samples can be identified by matching their pattern against those of known species of fish.

(b) DETERIORATION

The chemical and biochemical methods to be described are limited to measuring the extent of spoilage in the chilled state and of oxidative rancidity. Tests have been devised for other kinds of deterioration or for application to a narrow range of products but for one reason or another they have not been generally accepted and will not be described.

Three fairly well researched methods for measuring spoilage in the chilled state are available which depend upon the complex series of changes in flesh constituents brought about by autolytic enzymes and putrefactive micro-organisms. With certain precautions these methods can also be applied to products in addition to chilled fish in order to provide a measure of the amount of spoilage that has oc-

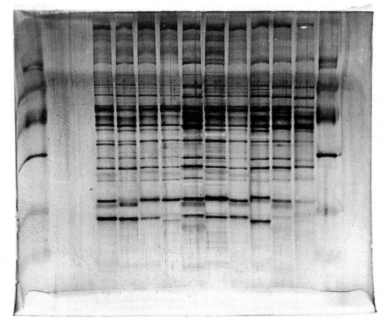

Fig. 6.6 The electrophoresis technique for identifying the species of unknown samples of fish results, as shown here, in a series of horizontal dyed bands of varying intensity. Each series is derived from the proteins extracted from different species of fish. The pattern of the bands is different for and characteristic of each species.

curred before processing. Thus, in principle, they are applicable to frozen, dried and canned fish. It should be noted that they do not necessarily tell us anything about wholesomeness; for this we have to turn to specific tests for pathogens. These spoilage changes result in the gradual accumulation in the flesh of compounds the quantity of which therefore provides a measure of the progress of spoilage that is independent of sensory assessment. The most well known of these compounds is trimethylamine (TMA) derived possibly partly by intrinsic enzymes but certainly by bacterial action from trimethylamine oxide (TMAO). Only marine species contain the latter compound so that the test does not apply to freshwater species. TMA, although the most prominent, is only one among several volatile basic compounds that increase in amount and an alternative method is to measure the total quantity of bases that can be easily volatilised (total volatile bases, TVB). The determination of TMA involves a fairly complicated chemical procedure; that of

Fig. 6.7 Apparatus designed by N. Antonacopoulos for the routine determination of total volatile bases in fish; the fish is boiled with mild alkali in the flask on the left and the vapours, containing the volatile bases, condensed on the right.

TVB, as shown in Figure 6.7, requires only a simple distillation followed by a titration of the condensed bases. Since TMA and TVB both contain nitrogen it is convenient to represent their concentrations in terms of milligrams of nitrogen in 100 grams of fish flesh. An entirely different method is based on the analysis of the breakdown product hypoxanthine (Hy) that similarly increases

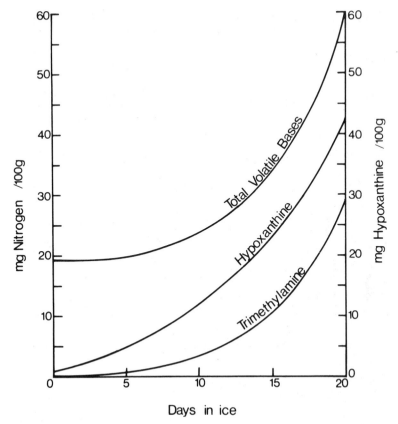

Fig. 6.8 The changes in concentrations of total volatile bases, hypoxanthine and trimethylamine with the degree of spoilage of cod; the exact form of the changes differs for different species.

in concentration in spoiling fish. The procedure for the determination of Hy is relatively complicated and requires some skill.

Figure 6.8 shows the general nature of these changes in cod. It will be noted that some TVB is present in even very fresh fish. In putrid raw fish held for 20–25 days in melting ice, the concentrations of TMA and TVB rise to about 50 and 70 mg nitrogen and that of Hy to about 50 mg/100 g fish, respectively. These values thus can be taken as the upper limits of what would be universally acceptable for this type of fish. For good or passable quality fish very much lower values would be set; for example, not more than 1.5 mg TMA nitrogen/100 g product has been recommended for very good quality cod for pre-packaging; 10–15 mg TMA nitrogen/ 100 g or 35–40 mg TVB nitrogen/100 g are usually regarded as the limits beyond which round, whole chilled fish can be considered too spoiled for most uses. In the case of other products different standards operate. Thus, not greater than 30 mg TVB nitrogen/ 100 g has been specified for frozen tuna and swordfish; not greater than 100–200 mg TVB nitrogen/100 g for a variety of salted and dried fish; not greater than 20 mg TVB nitrogen/100 g for the raw material used in various canned products. Caution should be exercised when comparing TMA values and especially TVB values because the result depends on the method of analysis.

The rate of increase of these compounds varies for different species so that if the spoilage of a test sample of any particular species is to be assessed, the particular rate in question must be known in advance. Various processes like freezing and canning do not affect these measures much, which thus affords means of assessing the freshness of raw material used in the products. However, before embarking on applying these chemical tests to any product one must be alert to the variations that the process may introduce in the values.

As in the case of spoilage judged sensorily, the change of these compounds with time after catching depends on temperature, the changes occurring more rapidly at higher temperatures. Thus from the measurement of concentration an estimate of time of storage at a given temperature can be obtained. This time is itself a measure of freshness. However, more importantly the concentrations also correlate reasonably well with the freshness as judged sensorily. That is, during storage under given conditions, changes in, say, odour go more or less hand in hand with changes in TMA concentration. From a measurement of concentration an estimate of quality or grade in terms of odour can thereby be obtained.

Since all these methods entail taking an extract or piece of fish they must destroy the sample, which may, nevertheless represent only a very small part of a batch. The methods involve fairly complex laboratory procedures lasting in general at least an hour. For this reason all are ruled out for assessing quality of chilled fish at port markets and for most factory reception areas. In the latter case they are used where good laboratory facilities are close at hand. They come into their own in two areas of quality control: for frozen or other stable products where time for analysis is not at a premium, and for providing a truly objective check on, or supplementary information to, sensory assessments of raw material. At the present time a few industrial and national specifications for salted, dried and canned products and for the supply of chilled raw material include a maximum concentration level of TMA or TVB which the values of the samples in a consignment must not exceed (the author is not aware of any instance where the relatively newly developed Hy method is being applied in this way). TMA and TVB are currently employed in some official regulatory laboratories as a means of checking or arbitrating on the quality of fish submitted to them.

In choosing which of the three methods to adopt various pros and cons come into play. As noted already TMA is restricted in application to marine species, but even here two important varieties, plaice (*Pleuronectes platessa*) and herring (*Clupea harengus*) and probably others not so far studied, do not form enough of the compound for it to be analytically useful. Also TMA does not increase much in concentration during the early stages of spoilage and its analysis requires a good deal of technical skill. TVB is of somewhat wider application and can be used for products not containing TMA (including some freshwater fish); its determination is relatively simple, cheap and rapid, does not require elaborate apparatus but it is not as specific as TMA and must be standardised if reproducible results are to be obtained. Hy can be used for more species and products (but not all) including several freshwater species and shellfish; unlike TMA and TVB it gives useful information about the degree of spoilage of very fresh fish; its determination is rather elaborate and requires skill. There is not a great deal of difference between the methods as far as variability between samples or relationship to other criteria of spoilage are concerned. In all cases results on a single sample must be interpreted with caution and properly representative numbers of samples analysed to obtain a valid measure of the average quality of a batch. Overall there is a marginal advantage in TVB for most industrial applications; for regulatory use and in arbitration Hy because of its better technical performance is preferred.

Turning to the chemical measurement of oxidative rancidity, there are two tests of roughly equal but limited value: peroxide value (PV) and thiobarbituric acid value (TBA). Oxidative rancidity, as occurs particularly in fatty fish, is a very complex deterioration in which oxygen first reacts with the unsaturated fats (lipids) to form hydroperoxides which then break down to substances that confer the objectionable rancid flavour. The PV is a measure of the first stage and TBA of the second, but unfortunately neither correlates well under all circumstances with sensory impressions of rancidity. The most that can be said is that if the PV is above 10–20 or TBA above 1–2 then the fish will in all probability smell and taste rancid. The measurement of PV depends on release of iodine from potassium iodide by the hydroperoxide and the titrimetric determination of the iodine; the units are the numbers of millilitres of 0.002 N sodium thiosulphate required to titrate the iodine liberated by 1 g fat extracted from the fish (this is equivalent to the numbers of micromoles of hydroperoxide). TBA is measured by a rather elaborate process, the units being micromoles malonaldehyde (one of the breakdown products) present in 1 g fat extracted from the sample. Both methods are used to a limited extent by both industrial and regulatory laboratories to check objectively that rancidity is not excessive.

A great deal of research has been done to find other chemical or biochemical methods for measuring deterioration in frozen and other processed fish but so far none have emerged that either are used industrially or can be recommended unequivocally. It seems unlikely that in the near future non-sensory methods will displace sensory methods from their dominant position in the fish industries.

3 Microbiological methods

The aim with these is to measure the numbers or to indicate the presence or absence of organisms in a fixed quantity of product. All the recommended methods of this kind depend upon comminuting the sample very finely in a suitable aqueous medium to release the organisms, diluting the suspension so formed and then, in the method still currently used in most laboratories, mixing it with a layer of agar jelly containing nutrients. When the layer ('plate') is incubated at a suitable constant temperature (between 20°C and 40°C), single organisms multiply into visible colonies that can be counted by the eye. The counts can then be related to a given unit weight of sample. Alternatively the sample is inoculated into a special medium that on incubation indicates only whether the organism is growing. This method gives no indication of the numbers of organisms but only whether they are present or absent in a given weight of sample. If a number of samples are examined

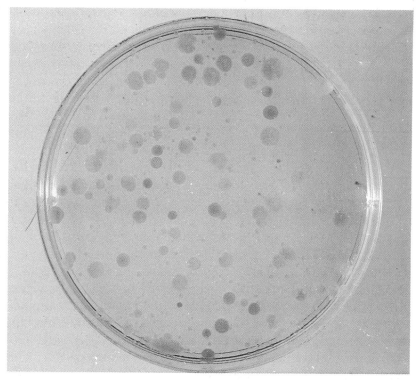

Fig. 6.9 One important form of bacteriological examination involves placing a known amount of the sample finely divided and suitably diluted on a nutrient jelly contained in a 'Petri' dish. Bacteria in the sample grow in the jelly as discrete 'colonies', shown here as nearly round patches of varying size. The number of colonies is then counted.

by this method the percentage showing a positive result can be obtained. For checking the degree of microbiological contamination of equipment, surfaces and workers, some means of swabbing or otherwise transferring the organisms to a sterile vehicle has to be adopted before plating can be done. Standard techniques for swabbing and sampling for microbiologically testing are described in texts on food hygiene.

Two types of method are useful for routine examinations: one that measures the total numbers of organisms present in the sample and capable of growing under the incubation conditions adopted, and those that measure the numbers of special groups of organisms, for instance, the pathogens. The first type is referred to as the Standard Plate Count (SPC) since it is conducted under generally recognised standard, uniform conditions. As SPC gives a comparative measure of the overall degree of microbiological contamination the interpretation of which has been discussed already; the temperature of incubation should be always quoted since the type and rate of growth critically depend on it. The second type utilise special media that favour nearly exclusively the growth of the particular group of organisms (e.g. *Escherichia coli*, *Staphylococcus aureus*) being measured. For the full enumeration of some groups more complex but essentially similar techniques are used. The results of microbiological examination are rather dependent on the exact method used. Thus, the details of sampling, comminution, diluting, plating out, composition of media, temperature and time of incubation, technique of counting, should all be kept standard. If a supplier is trying to meet standards set by a customer it is important that they both use identical methodology. Chapter 8 deals further with standard methods.

Plating and similar methods take 2–3 days to complete. This delay, as pointed out in Chapter 5, places severe limitations on their utility. In an effort to overcome this difficulty and also to simplify microbiological testing as a whole, a number of different methods are being introduced, all of which make use of automatic aids. The main development depends on the change in electric resistance caused to a nutrient medium when micro-organisms grow in it. This change can be detected at a very early stage in growth and can be related to the numbers of micro-organisms in the food sample introduced into the medium. These methods reduce the time to complete testing to a few hours. Also, they are cheaper and simpler to conduct than plating methods. It is likely that they will in time displace plating methods and provide information on which better decisions on microbiological quality can be taken.

Despite these advances, the equipment and skills necessary to use it are relatively costly. It is the case, therefore, that generally only larger companies and regulatory or public health laboratories have the necessary resources to carry out microbiological testing. Where it is necessary as a result of customer or regulatory demands for smaller companies to check occasionally the microbiological quality of their products, raw materials or water supplies, the use of a consultancy laboratory should be considered. The tasks of deciding which microbiological tests to carry out, where and how frequently they should be carried out demand specialist knowledge which should be sought before embarking on a programme of testing.

Microbiological quality control is applied routinely to many frozen and heat processed fish products, products eaten raw and occasionally to chilled products. Routine testing for sterility of canned products is almost always unnecessary; the process itself should be carried out so as to ensure an adequately low incidence of unsafe or spoiled containers. Adequate sterility results in complete softening of the bones of even large fish like salmon. Therefore bone softening offers a simple guide to the effectiveness of processing. If sensory inspection of contents should reveal any hardness, the batch should be examined for sterility. Tracing the causes of non-sterility in canned products entails special methods that need not concern us here.

4 Statistical methods

The most important part of quality control is process control. It can be achieved without much, if any, testing of product or knowledge of statistics. On some occasions, as already noted, it is essential to measure quality. Food quality cannot be properly measured unless the correct number, size and kind of sample is selected. To assess the freshness quality of one fish in the expectation that its measurement will give a valid concept of the quality of a tonne is clearly foolish because the quality of individual fish in the tonne will in general vary very considerably. To test every fish in such a batch in the hope of avoiding an incorrect measure is also out of the question except in the few cases where very large fish are being examined. How many fish to test in this kind of situation often can be worked out using statistical principles the details of which go beyond the scope of this book. It is not too much to say that measurements may be worthless without the application of a properly designed sampling scheme; if in doubt seek expert advice. Details of sampling and statistical methods are given in the further reading list.

Sampling inevitably entails a risk that the result of the measurement does not represent the true nature of the batch. The larger the number of samples the lower the risk. In circumstances where it is highly desirable that the level of risk be understood and at an appropriately low level, it is possible for the buyer and seller to agree on the level that is mutually acceptable. This level then fixes the numbers and kind of sample to be taken. When the penalty for taking undue risks is high, the numbers of samples taken for examination or the numbers of checks should be higher than in a less critical situation. Thus, if there is a risk to health from, for example, botulism the amount of inspection should be greater than that involved in examination for packaging defects. The number of samples taken for examination from a batch may have to be moderated by the costs involved. Costs arise from the actual labour of sampling and of examination but if the method used is destructive then also from the lost product.

Fish subjected to crushing and high temperatures at the bottom of a deep pile often will be poorer in quality than those at the top. In order to gain a correct impression of the overall quality of the pile, representative samples must be taken throughout the depth. In general, composition varies at different points in the same fish. Therefore, either correctly representative samples from all points of the body must be taken or the whole animal analysed. Similar examples of variability and the need to sample with caution occur throughout the fish industries.

7 Organisation of quality control and official inspection

There is no universally applicable organisation solution on how to achieve the maintenance and improvement of fish quality. The choice of the most appropriate system necessarily depends on a consideration of all the prevailing circumstances, but principally on the precise objective.

When the objective of the control is the protection of health then Governments in all countries recognise that they must assume ultimate responsibility and in almost all cases appropriate arrangements have been made. Thus, public health problems arising from the consumption of fish products are embraced by the national or local food laws, the enforcement of which falls in the first instance to a corps of official inspectors. The exact responsibility vested in these officials varies between countries but usually includes oversight of hygienic standards; it may include responsibility for ensuring absence of harmful entities like parasites, chemicals or pathogens. Some examples of official inspection will be given later. It is of course the responsibility of the fish industry in the first instance to ensure that harmful material is not consumed but official control for the removal of the worst quality is very widely applied.

Most Governments also assume responsibility for ensuring the operation of fair trading practices, which affect the fish industry in such matters as correct descriptions, labelling, weights and measures. Official inspection here takes the form of the examination of goods randomly sampled in shops, restaurants and the like, and of following up complaints from the public.

Whilst there is general agreement on these two activities the further extension of official inspection or involvement into what some would consider to be the province of industry is more controversial. Leaving aside occasions when they are themselves customers, should Governments concern themselves with such matters as regulating in industry eating quality, the form of presentation, the method of handling or manufacture? Attitudes differ on this question.

Some Governments and local authorities have assumed an interventionist role in this area, defining descriptions of different grade standards or minimum standards for human consumption and ensuring through the activities of inspectors that end-products comply with these standards. Such inspectors often have an additional role in advising and educating firms on how to achieve required standards. Sometimes the power of inspection extends to supervising and changing handling and processing conditions but as far as the author is aware, continuous in-plant official inspection of products is never required by Governments because it is prohibitively costly. In some instances officials act through voluntary agreement with the industry, in others they act mandatorily, that is with the backing of legal sanctions.

The arguments for having these types of official involvement are that: (a) in a fragmented undeveloped industry where there is no hope of change in a reasonable time, centralised direction whether mandatory or not is an effective (some would claim the only) means of rapidly improving technical efficiency and of reducing food wastage; (b) where a large, export industry earning valuable foreign currency is being set up or developed, it is vitally important to protect the quality of the commodity; in developing (and sometimes even developed) countries with many small firms entering such a trade, it is difficult for them individually to provide the level of technical expertise required; (c) the existence of an official system of checking and standardisation helps to create customer confidence and is likely to increase fish sales; (d) officials can combine work on health aspects with work on quality in general.

Other Governments and authorities contend that there is no, or at the most very little, need for centralised control over or concern about matters that are largely aesthetic and commercial. Normal market forces, in this view, should operate to regulate and eliminate poor quality. Objections to official involvement are: (a) it tends to be expensive; (b) once established it is difficult to demolish when the need for it has passed; (c) it is not responsive enough to changes in methods of handling and processing – it works with the system as it is; (d) it is a hidden form of subsidy; (e) it may delay the taking over by industry of those aspects of quality control that are really their responsibility.

Objective comparisons between these two attitudes is not easy

because it is never possible to compare exactly the total consequences or benefits of the same system with and without official
control. On the face of it there is not much difference between the
average quality of fish and fish products in countries like UK,
Belgium and France that have a policy of minimal intervention in
'commercial' aspects of fish quality with that in countries like USA,
Iceland and Canada with fishing industries subject to a relatively
much greater degree of official intervention. However, it is ad-
mitted that many different factors between these countries make
such a comparison suspect. Certainly, the protagonists of the first
attitude described can point to obvious successes of their policy
in circumstances like the rapid expansion of one sector of their
industry.

Among the world's fishing nations an impressive number operate
formal systems of official fish inspection and standardisation for all
quality attributes – which is a powerful argument in itself. It is also
true, as one writer has pointed out, that in regard to food quality
there is a tendency for Governments to become more paternalistic.
When considering ways of controlling quality and improving it,
probably the correct attitude is to review all possible approaches,
including those involving the various kinds of official intervention.
At the same time if an official system is adopted the purpose it is
serving and its effectiveness should be constantly reviewed. It is
commonplace to state that inspection for its own sake is worse than
useless.

Cutting across these two different attitudes is the acknowledge-
ment everywhere that the prime responsibility for turning out fish
products that are pleasing to eat and look at rests with industry. It
follows that the form of organisation of industrial quality control is
for it to decide. Much here depends upon the scale of operations.

Some firms may be so small that a formal organisation cannot
exist; it is only through the informal, part-time actions of, for
example, the proprietor, fisherman or employee that any inspection
or selection for quality is possible. In many instances this
arrangement works perfectly adequately provided the individual is
kept well-informed, is alert to customers' requirements and is able
and willing to correct things that are wrong or is able to put
innovations into practice. For this kind of situation a well-
organised, capable advisory service working close to the industry
is invaluable. It must be acknowledged that for understandable
reasons individual members of fish industries anywhere tend to be

conservative, suspicious of change, and are production – rather than product – orientated, so that instilling a concern for quality management in small firms can be an uphill task.

In large firms, cooperatives or associations, more opportunities for an organised system are available; in what follows this is the situation on which we will concentrate. If we define quality control as in Chapter 1 then there is a minority of fish processing firms at the present time where this concept is fully developed. However, there are many where a rudimentary or reasonable organisation already exists which could benefit even within their present size from the more rigorous application of the principles underlying the concept. Often the so-called 'quality control department', 'quality controller' or 'quality kitchen' fall short of being as effective as they could be. Their activities sometimes subsist in infrequent and sporadic checks of attributes unrelated to the qualities the customer or processor wants; sometimes their results are not examined or acted upon so that the feed-back essential to an effective quality control system is absent; often there is no clearly defined and fixed company or group policy with respect to acceptable levels of quality; often their subjective interpretations of quality change with the exigencies of the size and price of the raw material supply; they are sometimes divorced from or in conflict with those responsible for production.

In all these respects improvements are very likely to be possible. Fish is more variable in its intrinsic quality and subject to more variations than most other foods. Consequently there is in many fish industries a great need, currently still unmet, for systematic quality control systems. To provide such systems in an industry beset by uniquely difficult economic, environmental and social problems is a challenge. But it is a challenge that must be overcome if fish products are to compete effectively.

1 Industrial quality control

The basic principles of quality control organisation in the fish industry are little different to those applicable in any food commodity industry: the purpose here is to highlight aspects important to fish. First, there must be clearly understood specifications of what needs to be achieved in terms of end-product and process, and of what is required of the raw material. Second, the personnel whose responsibility it is to see that the specifications are met should have clearly defined functions and powers. Third, there must be sure means by which corrective action can be taken in the event of failure to meet the specifications.

(i) End-product specifications

The starting point in a well-regulated system is the preparation of a written specification of the end-product. Once this has been settled then it is possible to arrange the details of the rest of the system required to achieve the end-product. As pointed out in Chapter 1, this specification should be determined solely by customers' wants. The producer must therefore make every effort to find out those attributes of his product that are important from the market point of view. Sometimes the customer will provide enough information voluntarily to allow his needs to be specified, but otherwise the producer must rely on his own information or experience of the market. Attributes that come into consideration include freshness, species, amount of defects or deterioration, bacterial counts, ingredient formulation, quality and amount of additives, ingredients, incidence of bones, skin, parasites, appearance, portion or unit size and weight, form of packaging or protection, price, name. The amount of detail in a specification is dictated by the circumstances and the requirements of the customer.

Sellers or fishing vessel owners on an open auction market normally have only a general idea of what the buyers will bid for; written specifications are almost unknown in this situation. Similarly, a low volume, simple product going to undemanding, inarticulate customers requires only a very broadly-written specification. On the other hand, a high volume, complex, manufactured product destined for a demanding market consisting of consumers prepared to pay for it will probably require a fully itemised specification agreed in advance with customers. Specifications should ideally be precise in order to reduce the possibility of disputes arising over non-compliance. To this end as many objective descriptions of the attributes as possible should be included. If quality can be specified in terms of an exact grade designation (for example 1 or A) a description ('not sour'; 'no objectionable off-flavours') or a number (not more than 10 mg TMA nitrogen/100 g product'), so much the better. These devices are useful even though they may not be actually measured at any stage in production. If there is any argument about whether a specification has been met, the existence of a numerical value for it will clearly help the disputants to settle the matter. Dubiety can also be avoided if end-product specifications include descriptions of how testing is to be carried out. An example of a fairly comprehensive specification which is actually used in the purchase of fish for institutional caterers is reproduced in Appendix 1. Specifications can be set out in various ways though a regular form for them and for standards (Chapter 8) is emerging.

There must be a realistic expectation that specifications can be met over a reasonable period of time. Therefore they must take into account likely future variations in quality or properties. It is no use specifying high quality raw material if its continued supply is suspect. Moreover, details must be agreed with those responsible for production or line management; there must be no *ad hoc* changes in the specification as a result of temporary difficulties over the quality of the raw material or labour problems.

Most specifications remain within a firm or group of firms as guidelines for their own commercial use, but it is sometimes desirable to have specifications that are more widely agreed, nationally or internationally. Such specifications, which are commonly called standards, will be discussed in more detail in the next chapter.

(ii) Processing specifications and checking

Once the product is defined it is necessary to have a clear, agreed procedure for making it. Certain steps in handling and processing will be decisive for meeting the specifications; they should be noted and subjected to checking. Much of the contents of Chapters 3, 4 and 5 describing operations critical to the maintenance and improvement of quality is relevant to writing a process specification. Provisions for keeping temperatures low and times short should be prominent. Details of operations unrelated to fulfilling the product specification should be omitted. The kinds of points that could be included are:

methods of gutting, bleeding, washing
intimate mixing of fish and ice

ratio of fish to ice
period of stowage in ice or refrigerated sea water
maximum and minimum temperatures in chilled stowage
freezing time in the freezer and method of loading
temperature and time of cold storage
methods of glazing, wrapping, packaging
time and temperature of drying, smoking
period and temperature of brining
time and temperature of retorting
methods of cutting, filleting, skinning, deboning, trimming
avoidance of contamination with dirt, hair, scales, bones
hygiene, cleaning and sanitation measures
protective clothing and headgear

Product and process specifications are obviously closely inter-
related and occasionally it will be convenient to include in the
former details that might be placed in the latter. If not covered
elsewhere, it is worthwhile including in the process specification
information on the yields that should be aimed at. Although
perhaps more the responsibility of production or work study staff,
matters pertaining to yield and efficiency can conveniently be
combined with the main stream of quality control activities.

As well as describing correct methods of handling and so on the
process specification can include methods of checking, or can detail
points where inspection should take place. Its prime use is, of
course, as a guideline for process workers and supervisors whose
task is to put it into effect. Adequate staff of this kind should be
recruited and properly trained or instructed to carry out the amount
of checking required. It is impossible to lay down advice about the
numbers of staff and the exact nature of their tasks because individual circumstances vary so much. However, the importance of
proper control of handling and processing cannot be overstressed.
By good management and strict attention to detail at these stages
wasteful monitoring or possible discarding of finished product can
be avoided. Given the choice it is more important to put effort into
getting the process right than into extensive examination of products. Nowhere is this more true than in microbiological control –
it is far better to concentrate on hygiene and related matters than
on increasing the amount of testing. The value of fish increases
throughout processing and it becomes increasingly wasteful to
throw out or reprocess inferior material. Also, faults created at an

early stage are often impossible or at best difficult to eradicate.

Quality control personnel in the food industry generally pay a good deal of attention to what is known as the Hazard Analysis and Critical Control Point (HACCP) method. The HACCP method, whose title is self-explanatory, deals with similar principles of process control as described in this section; it deserves wider adoption in the fish industries.

(iii) Inspection of raw material and end product

Responsibility for the quality of raw material rests primarily with the supplier. The easiest way of dealing with poor quality raw material is to change one's supplier. Unfortunately, in the fish industries this kind of opportunity does not often occur and the buyer is usually forced to negotiate for better supplies at the next purchase. This process will normally involve both partners in a discussion about the nature of the poor quality in question; and possibly about how the faults can be rectified. Inevitably such a discussion will include methods for inspecting and judging quality.

Grading by fishermen or salesmen at the point of first sale of fish for species and size is generally accepted practice, but the amount of sorting for freshness and similar attributes is very variable. More often than not many aspects of quality must be assessed on the spot by the buyer or later by those responsible for quality in the factory. The methods, both sensory and non-sensory, that are available to help select raw material against a specification have been described in the previous chapter. In borderline cases involving differences of opinion over what is acceptable, it is useful to be able to fall back on a non-sensory method which then assumes the role of an impartial referee or arbitrator. Market conditions are sometimes such that batches of very variable quality have to be bought, in which case subsequent sorting for different end uses must be carried out. Normally, sorting for size and species presents no difficulty, but sorting for eating quality can be tedious, time-consuming and expensive; sorting for composition and comparable attributes is only rarely possible (for example pH or colour for large fish). With stable raw materials like dry ingredients, frozen or preserved fish, there is usually ample opportunity for any inspection or testing required.

Theoretically, if raw material is being selected properly, and handling and processing are carried out according to prescription, the end-product should always meet the specification. In practice mistakes occur and in general some end-product inspection and testing is a necessary safeguard. Where and how this occurs depends upon the kind of operation involved. Inspection for defects or some kinds of visually-apparent deterioration can be arranged at packing or wrapping points, or, in the case of continuous production lines, at waist-high conveyors. End-product inspection may also include continuous automatic checking for such aspects as weight and the presence of metal. Inspectors must be aware, of course, of the requirements of the end-product specification. Alternatively, representative samples from batches should be taken for laboratory or separate testing. The aspects of quality requiring examination are self-evident from the end-product specification; checking of aspects not required or expected by customers or legislation is unnecessary unless by doing so extra confidence in the product and improved sales can be engendered.

(iv) Recording, reporting and action

Keeping proper and adequate records and seeing they are acted upon are important functions which occasionally may be vital. The kinds of records depend upon circumstances but they should be selected with care. Measurements of critical aspects like bacterial counts, weights and yields clearly should be recorded; thermograph records of retorts and cold stores are usually regarded as very important. Unless they meet some useful purpose, are read or consulted quickly in order to reach a future decision, records and reports should not be made. Records should be clear and the information on them readily apparent; in particular, departures from acceptable values should be prominently indicated. The lines of responsibility for taking quality observations, for keeping and checking records, and for taking action should be clearly laid down in advance. Many useful ways of keeping and acting on records are available (control charts, inspection of trends, computer storage and retrieval); their use could with advantage be extended in various sectors of the fish industries. In particular, a growing variety of microelectronic devices and equipment is becoming available that can facilitate the work of quality control personnel in fish industries.

(v) Personnel and their responsibilities

Everyone taking part in getting fish from the sea (or site of primary production) to the consumer should have an awareness of the part they can play in maintaining or improving quality: every person should be his own quality controller. Everyone should be exhorted to submit ideas for improving quality: rewards for the introduction of successful ideas should be made. In large firms a special department responsible for organising and operating the manifold aspects of quality control can be set up. This department's role may be crucial in stimulating interest and improving control but obviously they cannot be everywhere at every time. Thus all should be encouraged to report to a responsible person departures from normal standards, mistakes in processing, accidents involving the product or its constituents.

The structural organisation of personnel specially responsible for quality control and their relationships to other staff will vary depending upon company policy. The arrangement should ensure that company decisions about quality are adhered to. Sometimes the aims of quality control and production staff differ, the latter tending to urge departures from standards in the interest of keeping up production. A mechanism for resolving difficulties of this kind should be set up, for example reporting to a higher level of management. Indeed the power of quality control staff to influence production should be clearly defined. Other important points in ensuring a successful quality control programme are: (a) the identification of management at all levels with the purposes and methods of the programme; (b) good liaison between quality control staff on the one hand and operatives, customers and suppliers on the other hand; in particular, means by which complaints from customers about products are received and rectifying action taken should be established; (c) a quality control staff alert to developments in fish technology, particularly those that have a bearing on quality maintenance and improvement or on the simplification of testing and inspection methods; (d) a quality control staff knowledgeable about the subtleties of fish quality, the factors that can affect it and the pitfalls that surround its assessment; (e) a quality control staff well-informed on the implications of food laws and other legislation.

(vi) *How much quality control?*

This is very difficult to answer, but the total cost of quality control should not exceed the benefits accruing from its operations. The trouble here is in evaluating benefits that are largely unquantifiable, for example the money value of improved quality. Another answer is that amount just sufficient to keep customer complaints at a low or zero level – a knowledge of which can only come from long experience of the market and after adjustments of the amount of quality control applied. A rule of thumb answer is that in this industry the bill should be about 1% of turnover. In view of the critical importance of achieving high quality for the success of a fish business, the expenditure of a sum as low as 1% of turnover is obviously money well spent. Within this guideline some flexibility must be allowed. Thus, as already mentioned the proportion of effort expended on a high priced, branded product is bound to be more in general than that on a basic, simple raw material. Another rule of thumb answer is to have as much as one's competitors.

2 Official inspection and requirements

We will here consider the ways in which some of the principal fishing nations (treated alphabetically) use laws or regulatory agencies in the task of controlling fish quality. The object is to be illustrative not comprehensive. Fuller details are available in specialist literature. As we will see, in some cases public bodies collaborate with the trade to provide jointly or by some other arrangement the service required; in others the public body assumes sole responsibility. For the most part, official arrangements of this kind are mandatory, that is industry must comply with statutory requirements; occasionally voluntary arrangements are offered to the industry.

The organisation, names and responsibilities of Government or local authority departments are apt to change and therefore the information cannot be guaranteed to be up to date in every respect.

(i) Canada

This country has probably the most highly-developed and extensive system of official inspection for fish products of any nation. All plants must be registered with and licensed by the Fisheries and Marine Service, Department of Fisheries and the Environment to process fish for any market. A condition of granting and retaining a licence is that the plant must reach certain minimum standards of construction and hygiene. Inspection for these purposes is carried out by a full-time staff of nearly 400 persons employed by the Inspection Branch of the Department.

The same corps of inspectors has powers to examine the standards of icing and stowage on board fishing vessels on arrival at ports and must grade all types of fish at unloading, in plant reception areas and on production lines into three freshness grades having well-defined features. Grade 3 (the lowest) is definitely spoiled, decomposed or otherwise unfit but is clearly of a higher standard than that generally considered condemnable in many countries: it is disposed of for fish meal or pet food. In the case of round, white gutted fish the quality at the lower boundary of this lowest grade corresponds to that after stowage in melting ice for 10–12 days. Uniformity of grading standards within the inspectorate is ensured by rigorous collaborative training. In the case of dispute between the firm and the inspector about the appropriate grade of a lot, it is held frozen under supervision. Frozen samples are drawn according to a sampling plan and sent to the Fish Inspection Laboratory for more detailed sensory and chemical analysis (TMA). The decision of the laboratory is final and appropriate action on the detained lot follows. By this means fish of the poorest quality is prevented from being marketed.

In addition, there is a voluntary programme of inspection under which Grade 1 products, that after official inspection and systematic examination comply also with certain prescribed standards of defects, can bear on the package the Canadian Government Specification Board's maple leaf insignia denoting guaranteed quality.

It is said that this official system does not supplant or render unnecessary control measures applied by the firm itself; the main objective is to monitor these measures and to provide technical advice where appropriate. One might question, however, whether the inspection process – now that it has achieved the recognition and establishment of improved quality grades – need continue at as high a level as before.

A somewhat similar compulsory system is in operation for the canned salmon industry using a relevant grading scheme. Two merchantable grades are in force: Grades A and B, the former being the superior pack. The grade must be stated on the label.

(ii) European Economic Community

The arrangements for inspecting fish and fish products in the 11 EEC countries are very varied and complex but consequent on the adoption of a common fisheries policy progress has been made on achieving harmonisation. There are, however, no community standards for fish or fish products.

At the present time inspection of chilled fish invariably occurs initially after or during laying out for auction at the main port markets. Depending on the country, inspection is of three kinds; (a) to ensure that fish unfit for human consumption is identified and disposed of; (b) to ensure good general standards of preservation and sorting on fishing vessels and at point of first sale; (c) increasingly, to supervise and control grading into defined categories of size and freshness. The first kind of inspection is carried out by public health officers in Ireland and the United Kingdom or by veterinary (or similarly qualified) officers in Belgium, Denmark, Greece, Italy, Federal Republic of Germany, France, Netherlands, Portugal and Spain. The regulatory authorities employing these officers in Belgium, Federal Republic of Germany, Greece, Netherlands, Portugal, Spain and UK are the local or regional Governments – mainly because in them are vested the powers of enforcing the state food laws pertaining to wholesomeness and safety. In Ireland, Denmark and France the State assumes overall responsibility through, respectively, the Department of Public Health, the Fisheries Inspection Service, backed up by the technical expertise of the centrally-located Inspection Service for Fish Products (Ministry of Fisheries) and the Veterinary Services Division (Ministry of Agriculture). The level of freshness quality at which condemnation occurs is similar throughout the EEC; it is roughly equivalent in the case of round, white gutted fish to stowage in melting ice for 14–16 days. Measurable differences in condemnation level are observable between ports where fish from distant waters, as distinct from near, is landed. Fish frozen at sea and chilled fish landed at smaller ports often escapes inspection of this kind. Examination before sale of shellfish for bacterial quality is usually the responsibility of the producer, though checking for compliance with recommended standards is carried out by the agencies just mentioned.

Mandatory inspection of the second kind is carried out in Ireland and Denmark by full-time officials of, respectively, the Fisheries Division (Department of Agriculture and Fisheries) and the Danish agencies just mentioned. Standards of icing, stowage, temperature control, hygiene and sorting are all covered.

The numbers of inspectors covering work of these two types varies from a few tens in Belgium and Ireland to about a hundred in Denmark, reflecting the different intensity of supervision deemed necessary in the 11 EEC countries having fisheries and the size of their fish industries.

Turning to the third kind of inspection, one part of the fish marketing policy of the EEC requires that chilled or iced fish of the main commercial varieties destined for human consumption should be offered at the point of first sale in defined categories of size (weight or length) and freshness. The standard sizes and freshness grades are meant to apply universally throughout the EEC but as noted above a small proportion of all landings is not graded. Responsibility for grading rests with fishermen, firms or associations but overall supervision and responsibility for initiating penalties in the case of infractions is in the hands of official inspectors.

Inspection of chilled and processed fish for public health aspects in factories, at inland markets, at retail and at port of entry in the case of imports is, in general, much less intensive. For example in the UK processing premises are visited only rather infrequently. Again, public health or veterinary officials employed by local or central Government are responsible for covering these sectors of the industry. In the City of London, authority for ensuring that fish unfit for human consumption does not enter into trade is uniquely vested in inspectors employed by a venerable trade association, the Worshipful Company of Fishmongers.

With the exception of Denmark, official inspection in factories and premises for general 'commercial' aspects of quality is not carried out. In Denmark, factories and other institutions handling fish must be approved by the Inspection Service for Fish Products and their products and processing methods regularly examined on the spot: processes (for example, temperature of frozen storage) must comply with regulations: any material not passing acceptable standards must be disposed of. The Danish Inspectorate uses as a basis for its actions the regulatory standards described in the next chapter. In France, factories manufacturing sterilised (canned and bottled) and semi-sterilised products must be regularly inspected for general cleanliness and standard of hygiene by officials of the Scientific and Technical Institute for Marine Fisheries (Ministry of Fisheries). In addition, representative samples of product batches are examined sensorily and analytically against mandatory standards of quality before permission to release the batches for sale is given.

Mandatory control of the temperature of chilled fish during all phases of distribution is enforced in Ireland by inspectors of the Department of Agriculture and Fisheries. The Governments of some EEC countries exercise a watchdog function over some special aspects of fish product quality, particularly those concerning compositional requirements. Thus, in the UK, Government departments sponsor the activity of the Food Advisory Committee whose function is to give advice to the government on a wide range of topics but including such matters as the minimum fish content of pastes and spreads, correct names of fish, and labelling requirements. The recommendations of this Committee usually result in legislation; however the amount of such legislation covering the fish industry is relatively small.

It will be appreciated from this survey that of the EEC countries, Denmark has the most highly developed system of official inspection and the UK probably the least. Nevertheless, governmental intervention aimed at regulating fish quality is increasing everywhere.

(iii) India

Rapid growth of a valuable export industry based on canned and frozen shrimp and prawn caused the Government to enact regulations requiring (a) certain minimum standards of cleanliness and hygiene in fishing vessels and factories handling shellfish for export; (b) mandatory factory control of quality using a set of defined standards; (c) mandatory pre-shipment inspection of representative samples drawn from production. The task of supervision, inspection, testing and certification is carried out by a team of trained Government personnel strategically disposed at the main processing centres. Their work is conducted under the aegis of the Central Inspection Agency, technical back-up being provided by the Central Institute of Fisheries Technology. In addition, support is provided by a Marine Products Development Agency.

As described in the next chapter, the Indian Government is also active through the Indian Standards Institution in drawing up official fish product standards. Products manufactured according to these standards and certified by the Institution may carry the ISI certification mark.

Apart from this, official regulatory activity towards fish quality is very fragmentary.

(iv) INFOFISH countries other than India

INFOFISH is an information and advisory service run under the aegis of the Food and Agriculture Organization (FAO) of the United Nations. It operates predominantly in the Indo-Pacific region, its member countries in addition to India being Bangladesh, Brunei, Hong Kong, Indonesia, Republic of Korea, Malaysia, Republic of Maldives, Pakistan, Papua New Guinea, Philippines, Singapore, Solomon Islands, Sri Lanka, Thailand and Tonga. The official quality control and inspection systems in these countries have been reviewed by INFOFISH in a report noted in the further reading list. The details in this report are too numerous to be effectively summarised here. In almost all cases the schemes in operation in INFOFISH countries apply only to products for export; products for home consumption are generally not covered.

(v) Japan

Mandatory inspection of chilled and frozen fish harvested locally or landed at Japanese ports from fishing vessels is carried out by a large corps of highly trained officials (generally veterinarians or pharmacologists) employed by the Food Inspection Services of the Health Bureau of the 46 prefectural Governments. The aspects included are (a) ensuring that fish are not so spoiled or otherwise affected (parasites, contaminants) as to be unfit for human consumption; (b) bacterial testing of raw shellfish; (c) ensuring that edible fish containing poisonous organs (for example, puffer) are identified, segregated and treated and despatched properly; (d) ensuring that adequate sanitary conditions obtain in the market and its surroundings. Testing for spoilage is done sensorily on representative samples; in doubtful cases samples are sometimes analysed chemically (TVB, TMA) and microbiologically in laboratories under the control of the Food Inspection Service adjacent to the main markets. Guidelines on sanitation, hygiene and microbiological standards are issued by the National Ministry of Health and Welfare.

Similar safeguarding of minimum safety and quality standards exist for inland markets, processors and retailers. Wholesalers must obtain a permit to trade from the Ministry of Agriculture and Forestry.

As far as control of eating or aesthetic quality is concerned, the only mandatory provision is that for exported frozen, canned and bottled products. Standards here are drawn up by the Ministry just mentioned but routine inspection is carried out by highly-qualified employees of two semi-Government agencies: the Japanese Frozen Foods Inspection Corporation and the Japanese Canned Foods Inspection Association. As the names imply these agencies are responsible for inspecting all exported foods of these kinds including fish. They are joint enterprises of government (Ministries of Agriculture and Forestry, of Health and Welfare, and of Trade and Industry) and the industries concerned. Inspection is based on taking representative samples of consignments and conducting in the factory in the first instance simple tasting and other tests. More elaborate testing is done in the agencies' own laboratories. The standard for frozen fish is a minimum single level that is considerably higher than the condemned level. Two standards for canned fish products are recognised: A (high class) and B (good commercial quality). No product that does not meet the requirements of the lower standard may be exported. The cost of inspection is covered by fees levied on the processors.

For the home market, further product standards have been drawn up by the Ministry of Agriculture and Forestry and a voluntary scheme of compliance with these standards is in existence for those processors wishing to take advantage of it. If the Ministry is satisfied that the producer can meet the standard consistently, the latter may for a fee use a special quality mark (Japanese Agricultural Standard). Regular inspection of these processors is done by the agencies just mentioned.

(vi) Norway

This country is one of the few where very specific requirements governing the handling and processing of fish are embodied in legislation. Various regulations lay down the exact way in which fish should be, for example, gutted, bled, washed, iced and stowed, dried, salted, frozen, cold stored and transported. There are mandatory rules for the quality grading of farmed salmonids. In addition, compulsory standards of construction of vessels and premises, of cleanliness, hygiene and sanitation are prescribed. Compliance in the case of all except canned fish is ensured by a large group of trained official inspectors under the General Directorate of Fisheries (Ministry of Fisheries); in the case of canned fish this responsibility is undertaken by the Norwegian Quality Control Institute for Canned Fish Products. Finished products are inspected by testing representative samples against minimum standards agreed with the appropriate sector of the industry. Processing can only be undertaken in factories approved by the Ministry's officials.

It is clear that Norway has a highly organised and comprehensive system of Government participation in fish quality control.

(vii) USA

The separate involvement of the State and National Federal authorities must be distinguished.

(a) STATE

Official responsibility for safety of raw shellfish and of some imported products rests in the first instance with the Public Health Departments of the State authorities. Comprehensive advice on recommended procedures is provided in manuals issued by the Federal Public Health Service (Department of Health, Education and Welfare). The State issues certificates to shellfish producers whose practices conform to these recommendations. Local officials also inspect chilled fish at landing for fitness for consumption, and, in conjunction with officials of the Federal Food and Drugs Administration, of factories for cleanliness and adequate sanitary conditions.

Two States, Maine and California, also collaborate with the local fish industries to operate mandatory full-time inspection programmes. The Maine scheme, administered by the State Department of Agriculture but funded by the industry, covers only factories making canned 'sardines'. It is almost entirely a finished-product scheme. Employees of the State located in each factory remove cans of product according to a routine sampling plan and despatch them to a central State testing laboratory where they are inspected for quality using standards agreed with the industry. Quality factors include can defects, fill and sensory properties.

The Californian scheme, administered by the State Department of Public Health, covers predominantly canned tuna but also mackerel and sardines. Every canning factory has a resident official inspector who supervises all aspects of cleaning, hygiene, raw material and ingredient quality, and processing. Also the finished product is examined in the factory against agreed standards of quality similar to those just mentioned for the Maine scheme. In both schemes lots are not released for sale until they are certified as having met the standards.

In the State of Maryland, compulsory regulations are in force that prescribe the permissible types of factory construction, of equipment and of processing operations for crab meat. Rather unusually a standard for bacterial count is also laid down.

(b) FEDERAL

The major involvement at the present time is by the National Marine Fisheries Service (Department of Commerce) in the formulation of product standards and specifications, and in administering a voluntary scheme of compliance based on these standards. Additionally, the FDA is engaged in the formulation of processing standards, in the inspection of imported products, and in the public health surveillance of processing establishments.

Voluntary US product grade standards, which will be discussed more fully in the next chapter, have been drawn up by NMFS fish technologists in conjunction with the industry and other interested parties, for most of the products moving in trade. Goods meeting the standard after inspection are allowed to bear one of four descriptions (Grade A, B, C or Sub-standard) depending upon which grade they fall into, or alternatively, 'Packed under continuous inspection'. In some standards Grade C is omitted. The use of these publically declared official guarantees of quality is said to create consumer confidence, stimulate trade in these commodities and place fish in better competition with other similarly-graded foods.

Compliance with product standards is ensured through on-the-spot examination by trained, full-time inspectors employed by NMFS. Samples are taken from product lots according to a plan and examined for the various quality aspects covered in the appropriate standard. As many as possible of the methods used in the examination are rapid, low cost and of high objectivity. The cost of inspection is recovered from the processor. The quality control systems operated by producers contracting for official inspection of this kind may be reviewed by NMFS staff, and if found to meet certain designated standards, given official approval. One objective is to reduce the amount of official involvement in quality control activities at processing plants.

Companies not wishing to have continuous in-plant inspection but nevertheless wanting lots certified as having reached these standards are able for a fee to send representative samples to an NMFS laboratory for examination.

The same Government agency has also been responsible for drawing up for a wide range of fish products federal specifications that lay down the requirements that must be met in goods purchased by all Federal and local authority agencies.

In the area of processing standards the FDA has drawn up guidelines on what are called Current Good Manufacturing Practice for food generally and for two fish products in particular. These will be discussed again in the chapter on codes of practice, but it is worth noting here that in a recent statement the FDA make it clear that certain provisions in these codes must be considered to be mandatory. The FDA also cooperates with the National Canners Association (a trade cooperative) in a national voluntary inspection scheme for salmon canning factories. Inspection by FDA or NCA officials covers fishing vessels, factories and all aspects of processing and finished product quality.

All this amounts to a very considerable engagement of US regulatory officialdom often against a background of acquiescence or even positive encouragement from the fish industries themselves.

8 Standards

Various meanings have been applied to the term 'standard' but here it is meant to refer to a reasonably complete and widely applied fish product specification that has been agreed nationally or internationally. Some kinds of official statutory regulations governing handling and processing practices are occasionally and confusingly referred to as standards; these are dealt with under codes of practice in the next chapter. Special statutory requirements in regard to composition (for example, colouring matters, additives, fish content) also are not included here. Fish product standards like most other food standards, are predominantly voluntary or recommended. The type of standard considered here is either (1) minimum, that is for all applicable quality attributes there are set out the lowest values acceptable in the product; or (2) multiple grade, that is there are set out the different groups of quality attributes that must be attained for the product to fall into one of two or more grades. Minimum standards represent the lowest common denominator of commercially and officially acceptable quality. The term regulatory is applied to standards used in conjunction with official inspection or certification schemes.

Standards serve three main ends: (a) to make clear the requirements that official authorities state are necessary – the authorities here act as watchdogs for consumers, often in collaboration with industry, (b) to facilitate trade by providing a commonly agreed basis for commercial transactions and so ensuring uniformity of approach, (c) to offer, it is hoped, a measure of consumer protection by removing from sale harmful or low quality goods.

The increasing number of standards for fish products, examples of which will follow, reflects a growing interest in and movement towards the standardisation of foods generally. Whether the considerable effort and expense involved in formulating fish standards and in ensuring that products comply with them matches the benefits associated with the three objectives is impossible to assess. Where statutory control of specific attributes of quality exists it is clearly essential that those operating the control should have a fixed set of standards to work to – the US and Canadian standards come into this category. Elsewhere, it is sometimes difficult to discern whether any pressing need exists for a standard. Experts recognise 'good' and 'bad' standards depending upon how clearly, exactly, completely and unambiguously they are written and upon how well they reflect commercial realities. The following brief descriptions of standards illustrates the range now operating in various countries: it is not meant to be comprehensive. An idea of the requirements and degree of detail set out in current standards can be gathered from Appendices 2, 3 and 4.

1 National standards

i Australia: has a very brief and basic voluntary standard for general categories of product; it is produced by the National Health and Medical Research Council.

ii Canada: has a comprehensive set of mandatory standards for most commercial fish products. These are very detailed, usually with two acceptable grades. They are drawn up by the Inspection Branch, Fisheries and Marine Service (Department of the Environment) in connection with the scheme of official inspection described in the previous chapter.

iii Denmark: to serve the requirements of the Fisheries Inspection Service a number of mandatory, comprehensive three-grade standards are available for the main commercial products. Responsibility for the compilation rests with the Danish Quality Committee on which are represented the Government officials and the industry.

iv France: the Scientific and Technological Institute for Marine Fisheries has drawn up detailed, mandatory standards for the canned and semi-preserved products covered by their inspection service.

v German Democratic Republic: has over 20 national standards of the kind considered here. These are comprehensive, usually two- or three-grade standards used by the Technical Control Organisation. The TKO, working under the aegis of the State Office for Weights and Measures, is responsible for quality control in State factories.

vi Federal Republic of Germany: the Federal Association of the

Fish Processing Industry has issued for a number of products moderately detailed voluntary minimum standards which are used as a basis for intercompany dealings. In addition, the German Agricultural Association has drawn up schemes by which fish products can be graded according to scores given to various quality factors.

vii India: the Indian Standards Institution has produced about 30 voluntary standards covering products of particular interest to that country. These standards are minimum but detailed. One dealing with crab meat canned in brine is shown in Appendix 2.

viii Ireland: fully comprehensive voluntary minimum standards for 11 frozen fish products have been published by the Institute for Industrial Research and Standards. That for frozen smoked demersal fish fillets is shown in Appendix 3.

ix Japan: for the products with which this book is concerned two sets of standards are in operation. The first are detailed mandatory standards issued by the Ministry of Agriculture and Forestry in conjunction with compulsory inspection of exported canned and frozen products. The canned are two-grade and the frozen are minimum standards. The second set comprise Japanese Agricultural Standards issued by the same Ministry for voluntary use within the country. These are, in fact, similar in scope and character to the export standards.

x Norway: has fairly detailed two-grade mandatory standards for 15 or so canned products; they are used as the basis for inspection by the Quality Control Institute for Canned Fish Products.

xi UK: the semi-Government agency, the Sea Fish Industry Authority, has published model, detailed, minimum standards for a range of chilled and frozen products to assist local authority procurement officers.

xii USA: has two types of standard: Federal and a single one drawn up cooperatively by the 'sardine' canners in Maine. The former comprise (a) fully comprehensive two- or three-grade voluntary standards formulated by the National Marine Fisheries Service in support of their system of voluntary inspection, (b) 'specifications' for a wide range of products drawn up in similar detail by the Service for the benefit of official procurement agencies. Part of the US Government Federal Register describing the grade standards for whole or dressed fish is reproduced in Appendix 4. The Maine standard is used in the scheme of official inspection described earlier. It is a three-grade standard of considerable detail, and it is interesting to note that it involves, in part, weighting the scores given to different sensory attributes.

xiii USSR: the full texts in Russian of the standards for fish and fish products used in this country are given in a book published in 1967 (Izd. Komiteta Standartov, Mer i Izmeritel'nykh Priborov Pri Sovete Ministrov SSR).

2 International standards

i European Economic Community: a mandatory regulation, mentioned in the previous chapter, controls the grading system at first sale of chilled fish. The fish must be sorted identifiably before sale into thee freshness grades, and depending upon the species, into several size grades. The regulation gives details of how the grading classification is to be carried out; otherwise it is less detailed than many product specifications.

ii Joint FAO/WHO Food Standards Programme – Codex Alimentarius: Currently about 34 countries are collaborating in the drafting of comprehensive, minimum standards for a wide range of fish products moving in international trade. Almost all are for products meant for direct sale to the consumer. Those dealing with Frozen Pacific Salmon, Frozen Blocks of Fillets and Frozen Fish Fingers and Portions cover products which are meant for further processing.

At the time of writing 13 recommended standards have reached step 9 of the lengthy 10-step procedure of discussion between member Governments and interested agencies. At this point the standard is published and submitted to Governments for their formal acceptance. So far 40 countries have accepted some measure of acceptance of these standards. The assumption is that standards acceptable on a worldwide basis and published in the Codex Alimentarius will be legally binding in those countries operating them. It has been possible to reach a common view on matters such as hygiene, contaminants, exact specification of intrinsic defects and of defects of workmanship, but the task of agreeing objective measures for other important attributes has defeated the drafters and the technologists. Thus, none of the standards specifies in numerical terms acceptable degrees of staleness, chemical deterioration or microbiological contamination.

3 Microbiological standards

As stated already the numbers of different organisms in a given quantity of product provides information on two main points – the contamination with pathogens, and general state of cleanliness or spoilage. It would therefore seem logical to want to specify in a quality standard the maximum numbers of organisms that should be permitted. For some products such as raw shellfish where there is a direct relationship between the risk to health and incidence of harmful organisms the force of this argument is universally accepted. In these cases numerical microbiological standards are commonplace and as was pointed out in Chapter 2, are applied compulsorily by local and national authorities. Even in the case of products where the connection between health risk and count is tenuous, for example those eaten only after cooking, companies have judged it wise to establish upper limits for counts beyond which the goods are not accepted for purchase. These limits are not infrequently written into intercompany specifications where they act as an incentive for the supplier to adopt proper sanitary conditions in the preparation of his wares.

On the other hand, when we come to national or international standards some of which might be mandatory, experts differ as to the need or usefulness of specifying counts in products in general. Opponents of inclusion point to the lack of agreed methods of analysis and sampling, the difficulties of interpreting measurements of microbiological incidence, and because of the inhomogeneity of microbiological occurrence, to the uncertainty of detecting potentially harmful samples in a production lot. It is particularly unwise in their view to lay down standards for total undifferentiated plate count because this measurement gives only a very general idea of either the sanitary condition under which the food was prepared or the freshness. In the face of these difficulties considerable problems of enforcement arise. Proponents of inclusion are confident these problems are being or will be overcome or minimised. A number of national regulatory authorities do apply indicative but strictly enforced microbiological standards to fish products imported into their countries. Although similar, the numerical values set out in these standards for the incidence of different pathogens in the same produce are not identical. Of particular relevance to this discussion are the findings of the International Commission on Microbiological Specifications for Foods. This body of experts has reviewed and published (see further reading list) values for SPC and pathogens, commonly used in different countries for different fish products. The main results are shown in Appendix 5; they provide essential guidelines for regulatory bodies and companies, and are the starting point for any agreement to introduce microbiological values into fish standards.

4 Standard methods of analysis and sampling

Whenever a limiting numerical value applied to an attribute is quoted in a standard it is often highly desirable to include alongside, a single, exact and fully detailed method of analysis. This is because different methods of analysis are liable to give somewhat different results with many food attributes like microbiological count, drained weight, amount of glaze, TMA, content of fish, chlorinated hydrocarbons. The specification of a so-called referee, reference or standard method of analysis that must be used to check compliance with the value overcomes this difficulty. There is no objection to the referee method being complex or cumbersome because in the main it would only be used infrequently as for example in cases of dispute. It will be noted that some of the standards referred to in this chapter contain such methods.

The validity of a measurement on a lot of produce depends upon the method of sampling. Thus it is desirable wherever possible that the method be precisely specified. If the sampling plan laid down is followed whenever withdrawals take place measurements taken on different lots can be reasonably expected to have the same probability of being valid. Standards current in Canada, Ireland, India, Japan and USA all include sampling plans. Even for the same product, the plans differ somewhat, suggesting that different assumptions have been made in the calculations. The subject of technically correct and harmonious sampling plans is part of the international Codex Alimentarius programme; one sampling plan has been produced so far (Prepackaged Foods – AQL 6.5; CAC/ RM42-1969). The only fully tested and nationally agreed compilation of standard analytical methods for fish and fish products is to be found in Section 18 of the manual of the US Association of Official Analytical Chemists; methods include drained weight, net contents, fish identification, TMA.

International agreement on methods of analysis of fish products also has made little headway. The inclusion of chemical methods of measuring spoilage was considered for Codex fish standards but rejected, at least for the time being, because of technical disagreements. As referred to in the previous section, progress towards international agreement on microbiological methods of enumeration and on related methods of sampling has been made by the International Commission on Microbiological Specifications for Foods.

9 Codes of practice

As with the term 'standard', various interpretations are placed on 'code of practice'. Here it is meant to be essentially a processing specification that applies generally to processes of a given type. An industrial processing specification deals with the particular product or equipment that is of concern to the individual firm. It may for example include details of how a particular make of machine should be operated. A code of practice on the other hand ranges more widely usually without reference to particular styles of product or types of process. Codes are therefore guidelines normally for a whole industry drawn up by regional, national or international agencies. They are almost invariably voluntary or advisory in nature, largely because it would constrain industry unduly to make products in enforced, exactly prescribed ways. Occasionally, some of their features, particularly those relating to health protection, have the force of law – for example the US FDA's Good Manufacturing Practices referred to later in this chapter. Moreover, processing details are included in some mandatory standards; for example, the Codex Recommended International Standards for Quick Frozen Fillets specify how the freezing shall be carried out. These confusions have led some authors to include codes of practice with standards.

In form they are compilations of discrete pieces of advice or sets of recommendations based on accumulated practical experience of how to achieve good products. Principles are in general eschewed, though they may be included to explain a point. Sometimes they are simple lists of do's and don'ts: recent examples tend to be very comprehensive – almost handbooks or manuals. Their usefulness for practical quality control rests in their succinctness and common-sense directness. An ideal code should leave as little as possible to interpretation or to the imagination; by following it to the letter success should be assured. However, codes do need revision from time to time because new ways of making mistakes are always being uncovered and new processes are being continually introduced. The following is a selection of prominent examples.

1 National codes

Australia has two codes of relevance to the fish industry – one for the production and handling of frozen products (National Health and Medical Research Council) and another for the handling and processing of abalone (Department of Primary Industry). Both are technological in emphasis; hygiene aspects receive only passing attention.

The German Democratic Republic has a long series of codes and recommendations published as specialist sector standards (TGL – standards beginning with the prefix 81).

The Indian Standards Institution has available two codes related specifically to public health aspects: (a) Code for sanitary conditions, handling and transport to fish industry (IS 4303 – 1967) (b) Recommendations for maintenance of cleanliness in fish industry (IS 5735 – 1970).

The Standards Association of New Zealand has produced a code relating to all aspects of handling, processing and distribution (NZS 8402:1974). This code includes recommended temperatures and times, and hygienic requirements as well as names and illustrations of food fish marketed in the country. The New Zealand Fishing Industry Board has published codes for eel and mussel processing and for the air freight of chilled fish.

In the UK two codes published jointly by the Ministry of Agriculture, Fisheries and Food, and the Department of Health and Social Security relate to hygiene in the retail industry and in transport and handling of fish; a third deals with the preparation and handling of hot smoked trout in order to eliminate the risk from botulism. These are complemented by the British Standards Institutions 'Recommendations on cleaning in the fish industry' (BS 4259:1968). A trade association (UK Association of Frozen Food Producers) has published a code giving recommendations for the handling, production, distribution and retailing of frozen food much of which is relevant to the frozen fish industry. The Sea Fish Industry Authority has published guidelines for the handling of chilled fish by retailers, and for fish packed in controlled (modified) atmosphere.

Although not named as such the recommendations and regulations of the US FDA in relation to sanitary preparation of food are in effect codes of practice. These are published under the Code of Federal Regulations, Part 128, Title 21 – Human Foods: Current Good Manufacturing Practice (Sanitation) in manufacturing, processing, packing or holding. This main article relates to foods generally but two subparts concern especially smoked and smoke-flavoured fish and frozen, raw, breaded shrimp. As pointed out above, parts of these are mandatory in contrast to all the other codes so far described, which are entirely voluntary. A rather similar mandatory code is embodied in a 1964 ordinance (735, p. 1, sec. 70–55) enacted by the City of Milwaukee. In this a method for the safe preparation of smoked fish is laid down. The need for such a code arose after an outbreak of botulism caused by eating smoked fresh-water fish (ciscoe and chubb) prepared in the region of the municipality. Two voluntary codes have been issued by national organisations representing the interests of the frozen food industry of the US. They are both quite comprehensive and in major part are applicable to the frozen fish industry: 'Frozen Food Handling Code' (Association of Food and Drug Officials, 1961) and 'Voluntary operating practices' (Frozen Foods All-industry Co-ordinating Committee 1961).

2 International codes

The International Institute of Refrigeration (IIR) have published sets of recommendations for the processing and handling of frozen foods, and for the chilled storage of perishable produce, both of which cover fish products.

Undoubtedly the most comprehensive and widely applicable codes are those compiled by the FAO Department of Fisheries. Principally these voluntary codes are meant to provide technical guidance to manufacturers wishing to make products that meet Codex Alimentarius Standards (or similar codes), and therefore in 'draft' form they are being subjected to the same vetting procedure afforded the Standards themselves. However, in fact they are virtually finished pieces of work and could be used with advantage as they stand. So far 11 codes have been prepared – for frozen fish, fresh fish, canned fishery products, smoked fish, and shrimps and prawns, lobsters and related species, salted fish, minced fish blocks, crabs; frozen, breaded and battered fishery products; hygiene for handling molluscs. They should be read in conjunction with the 1969 FAO/WHO Recommended International Code of Practice – General Principles of Food Hygiene. The most recent draft revised codes – for frozen fish excluding shellfish and precooked fish, fresh fish and canned fish – incorporate relevant details from this hygiene code and are thus complete in themselves. The frozen fish code lists nearly 200 recommendations. An excerpt showing the style and detail is given in Appendix 6. Like so many other codes and standards FAO codes are based predominantly on northern hemisphere practical experience, though much of their advice should be applicable to fish industries anywhere.

Appendix 1
Model purchase specification

FROZEN FILLETS OF WHITE FISH

1 Scope of this specification

This specification applies to whole fillets purchased frozen, of the following species:

Cod (*Gadus morhua*)
Haddock (*Melanogrammus aeglefinus*)
Whiting (*Merlangius merlangus*)
Plaice (*Pleuronectes platessa*)
Lemon sole (*Microstomus kitt*)
Saithe (*Pollachius virens*)

Redfish (*Sebastes* sp.)

South Atlantic hake (*Merluccius* sp.)
Greenland halibut (*Reinhardtius hippoglossoides*)
Dover sole (*Solea solea*)

2 Definition of fillet

Fillets are slices of fish muscle which have been removed from the carcass by cuts made parallel to the backbone and from which all internal organs, head, fins, bones (other than pin bones) and substantially all discoloured flesh have been removed.

3 Specification of product as delivered

The fish on delivery shall meet all of the following requirements:

(i) Colour (blemishes)

Bruises and blood clots and other localised discoloration which materially affect the appearance of the fish and/or eating quality shall be absent.

(ii) Bones, skin and belly lining/flap

The fish shall be free of bones except pin bones. The fish shall be supplied **skin-on**/skinless. **(Fish which is deemed skinless should contain no more than 10 cm^2 of skin per 3.2 kg (7 lb) unit.)** No more than 15 cm^2 of belly lining shall be permitted per 3.2 kg unit. An excessive amount of belly flap must not be present and should constitute not more than 10% of the total fillet weight.

(iii) Worms and other parasites

The maximum tolerance for nematode worms is three worms per 3.2 kg (7 lb) unit. No other parasites are permitted.

(iv) Size of fillets

Individual fillets must be not less than g (oz), **or more than kg (lb)** each in weight. (Insert required weights.)

(v) Eating quality

The fillets must not gape or exhibit abnormal textural faults (e.g. due to the use of spent or starved fish) nor must they contain abnormal intrinsic odours or flavours such as 'weedy' or 'diesel' flavours common in fish from certain grounds at certain seasons.

   At the time of delivery the fish must meet the following freshness standard:

(a) A minimum score of 6 in the taste panel system for assessing freshness.
(b) The fillets on cooking must be free of the following:

1 Objectionable cold-storage odours and flavours.
2 Toughness and dryness resulting from cold-storage deterioration.
3 Gelatinous texture.

(c) In the case of cod and haddock the values indicated below on the taste panel system for assessing cold-storage deterioration must not be exceeded.

Cold-storage flavour	3.0
Firmness	4.5
Dryness	3.0

(vi) Packaging

The frozen fish fillets must be wrapped so that they are protected from bacteriological and other contamination, and from dehydration during storage and transportation.

Any container used for the delivery of frozen fish fillets must be constructed of such material that taint is not imparted to the fillets.

Fillets should be packaged for delivery in units of kg (lb).

(vii) Temperature at delivery

The temperature of the frozen fish fillets at the time of delivery must be no higher than −15°C (5°F).

(viii) Freezing and cold-storage

Freezing must be carried out according to the code of practice for quick freezing. Storage prior to delivery should be at −18°C (0°F) for no longer than 3 months. Longer periods of storage must be at −29°C (−20°F). Cold-storage temperatures must not be allowed to fluctuate.

Bold type in Section c of this specification indicates that options are to be exercised by purchasers.

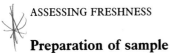

ASSESSING FRESHNESS

Preparation of sample

Fish to be tested must be steamed in a closed dish over boiling water for 35 minutes or, if frozen, for 18 minutes after thawing. The dish should remain covered and be kept in a water bath at 60°C during testing.

Scoring system; cod, haddock, whiting, saithe, south Atlantic hake

Fresh, sweet flavours characteristic of the species	10
Some loss of sweetness	9
Slight sweetness and loss of the flavour characteristic of the species	8
Neutral flavour, definite loss of flavour but no 'off' flavours	7
Absolutely no flavour, as if chewing cotton wool	6
Trace of 'off' flavours, some sourness but no bitterness	5
Some 'off' flavours and some bitterness	4
Strong bitter flavours rubber-like flavour, slight sulphide-like flavours	3
Strong bitterness but not nauseating	1
Strong 'off' flavours of sulphides, putrid, tasted with difficulty	0

Scoring system, Redfish

Very sweet or intense sweetness, characteristic flavour	10
Definitely sweet, nutty (Brazil), meaty	9
Loss of sweetness, nutty (Brazil)	8
Very slight sweetness, neutral, milky, chestnut, slight off nut-oil	7
No sweetness, no sourness, tasteless, cotton wool	6
'Off' flavours, bad nuts, herring, some sourness, slight rancidity	5
Strong 'off' flavours, strong sourness, some bitterness, rancid oil	4

Strong bitterness, biting rancidity, sulphide flavours   3
Nauseating flavours, difficult to hold in the mouth   2
Nauseating, putrid   1

Scoring system, Plaice

Meaty, shellfishy, earthy   9
Sweet, and meaty, (or oily fresh herring-like)   8
Sweet, meaty with curry, peppery or spice flavour   7
Neutral or bland   6
Rancid, slightly sour   5
Sour and bitter   4
Strong sour and strong bitter, rotten fruit   3
Nauseating   1

ASSESSING THE COLD-STORAGE DETERIORATION OF COD AND
HADDOCK

Preparation of sample: As for freshness assessment

| | Texture | |
Cold Store Flavour	Firmness	Dryness
0 Absent	0 Very soft	0 Sloppy, watery
1 Very slight	1 Softer than normal	1 Juicy (normal fresh unfrozen North Sea cod)
2 Slight	2 Firm (normal 2–5 days iced unfrozen fish)	2 Slightly dry
3 Moderate	3 Slightly firmer than normal	3 Dry
4 Strong	4 Slightly tough	4 Extremely dry
5 Very strong	5 Tough	
	6 Extremely tough	

Appendix 2
Indian Standard specification for crab meat canned in brine

Foreword

This Indian Standard was adopted by the Indian Standards Institution on 24 December 1973, after the draft finalised by the Fish and Fisheries Products Sectional Committee had been approved by the Agricultural and Food Products Division Council.

Canning of crab meat has been started .in the country on a moderate scale for export as well as internal consumption. It is hoped that the formulation of an Indian Standard on the subject would help in processing and canning of good quality crab meat in the country under hygienic conditions.

Crab meat is obtained from fresh crabs. The crabs are washed, cooked and deshelled and crab meat after blanching is canned.

For the purpose of deciding whether a particular requirement of this standard is complied with, the final value, observed or calculated, expressing the result of a test or analysis, shall be rounded off in accordance with IS: 2-1960. The number of significant places retained in the rounded off value should be the same as that of the specified value in this standard.

1 Scope
1.1 This standard prescribes requirements and methods of sampling and test for crab meat canned in brine.
1.1.1 The term crab shall apply to the edible species of the genera *Scylla* and *Protunus*.

2 Terminology
2.1 Blanching – Heating the crab meat for an adequate period so that it attains the characteristic flavour and texture.

3 Requirements

3.1 Hygienic requirements – The material shall be prepared, filled and processed under hygienic conditions and only in premises maintained in a thoroughly clean and hygienic maner [*see* IS: 4303 (Part II)-1967].

3.2 Raw material

3.2.1 The raw material used for preparation of canned cab meat shall be fresh crabs without any noticeable injury.

3.2.2 Vacuum-dried salt or common salt conforming to IS: 594-1962 shall be used.

3.3 Preparation and processing

3.3.1 Species, such as *Scylla serrata*, which can live for a considerable time outside water, shall be paralysed instantly by putting them in ice water These crabs shall be packed in ice straight away.

3.3.1.2 The crabs packed in ice as mentioned above are dressed by removing the dorsal, viscera and gills. The material shall be washed in potable water jets to remove slime and dirt.

3.3.1.3 The crabs so cleaned shall be precooked for sufficient time. After cooling, the meat shall be separated. The claw meat and body meat shall be kept separately.

3.3.2 The claw meat and body meat separated after deshelling shall be properly blanched in boiling brine containing citric acid, if desired. The blanched meat shall be cooled, and filled in sulphur-resistant lacquered cans lined with parchment paper. The packed cans shall be filled with brine of sufficient strength to get optimum salt concentration, mixed with adequate quantity of acetic acid and citric acid to prevent bluing and blackening.

3.3.3 The cans shall be exhausted by heat, steam or mechanical process and sealed in hot condition by double seaming. The sealed cans shall be processed at such temperature and for such length of time as will ensure adequate sterilisation of the finished product without burning, scorching or overcooking. Water used for the cooling of cans shall be maintained in clean condition and shall be chlorinated to maintain residual chlorine concentration of 5 ppm. The meat shall be packed with claw meat sandwiched between body meat.

3.3.3.1 The can exterior, especially seams, shall be free from dents, rust, perforations and distortions. The cans shall not show leaking, panelling or swelling. The interior of the can on opening shall not show any visible black discoloration, rusting or pitting and the inside lacquer shall be in good condition.

3.3.4 The packed can shall be adequately filled with brine of sufficient strength so that optimum salt concentration is obtained in the product. The filled cans shall be exhausted, seamed, washed and processed in steam under pressure. The heat treatment shall be for such a length of time and at such a temperature that there is no overcooking and at the same time, the cans are made commercially sterile. The heat treated cans shall be cooled in water chlorinated to 5 ppm available chlorine.

4 Requirements for finished product

4.1 The contents of the can on opening shall present a characteristic colour and odour of crab meat and shall not give any foreign odour.

4.2 The material shall be free from scorched, bitter or any objectionable flavour.

4.3 The material shall be free from stains, dirt, insect or hair or other extraneous matter. It shall be free from veins, membrane, shell particles and pieces of appendages.

4.4 The material shall be free from bluish colour.

4.5 The material shall be free from any poisonous and deleterious substances.

4.6 Preservatives – The material may contain the preservatives and firming agents permitted under Prevention of Food Adulteration Rules.

4.7 Drained mass of the contents – The drained mass of the contents in each can shall be not less than 65% of the net water capacity of the can as tested by the method given in Appendix B of IS: 2236-1968.

4.8 The material shall also conform to the requirements prescribed in Tables 1 and 2.

5 Packing and marking

5.1 Packing

5.1.1 *Packing in cans* – The material shall be packed in cans uniformly coated internally with sulphur-resistant lacquer and lined with parchment paper. These shall be sealed hermetically. The lacquer used shall be such that it does not impart any foreign unpleasant taste and smell to the contents of the can and does not

Table 1 Requirements for crab meat canned in brine (*Clause* 4.8).

SL No. (1)	Characteristic (2)	Requirement (3)	Method of test, Ref. to Appendix of		
			This Standard (4)	IS:2168-1971[1] (5)	IS:2236-1968[2] (6)
i)	Vacuum in can in mm, *Min*	150	—	—	A
ii)	Sodium chloride in brine, per cent by mass, *Max*	2	—	—	C
iii)	Acidity of brine as citric acid (anhydrous) per cent (m/v), *Max*	0.2	—	—	D
iv)	Bacteriological requirements	To satisfy the test	—	G	—
v)	Acid insoluble ash, per cent by mass, *Max*	2	A	—	—

Table 2 Limits of metallic impurities in crab meat canned in brine (*Clause* 4.8).

SL No. (1)	Characteristic (2)	Requirement (3)	Method of test, Ref. to Appendix of of IS: 2168-1971[1] (4)
i)	Arsenic, ppm, *Max*	1	B
ii)	Lead, ppm, *Max*	5	C
iii)	Copper, ppm, *Max*	10	D
iv)	Zinc, ppm, *Max*	50	E
v)	Tin, ppm, *Max*	250	F

[1] Specification for pomfret canned in oil (*first revision*).
[2] Specification for prawns/shrimp canned in brine (*first revision*).

peel off during processing and storage. The lacquer shall not be soluble in brine to any extent.

5.1.2 The cans may also be lacquered externally subject to agreement between the purchaser and the vendor.

5.1.3 *Packing in cases* – Unless agreed otherwise between the purchaser and the vendor, the cans shall be packed in cases, strong enough to withstand rough handling by rail, road or sea-transport without damage to the contents. The number of cans in each case shall be as agreed to between the purchaser and the vendor.

5.2 Marking – The labelling of the cans shall be done either by printing or lithographing on the cans themselves or by attaching a label, subject to agreement between the purchaser and the vendor.

5.2.1 The can and label together shall give this information:

(a) Name of the material with the brand name, if any:
(b) Name and address of the manufacturer (optional for export purposes);
(c) Minimum net mass or the drained mass of the contents of the can in grams (and also in ounces, if required by the purchaser, optional for export purposes);
(d) Batch or lot number and the date of manufacture in code to be embossed on the can;
(e) List of additives added; and
(f) Licence number, if any, under which the manufacturer has been permitted to can the material.

5.2.2 The warranty period may also be mentioned on the label subject to agreement between the purchaser and the vendor.

5.2.3 Each container may also be marked with the ISI Certification Mark.

Note – The use of the ISI Certification Mark is governed by the provision of the Indian Standards Institution (Certification Marks) Act and the Rules and Regulations made thereunder. The ISI Mark on products covered by an Indian Standard conveys the assurance that they have been produced to comply with the requirements of that standard under a well-defined system of inspection, testing and quality control which is devised and supervised by ISI and operated by the producer. ISI marked products are continuously checked by ISI for conformity to that standard as a further safeguard. Details of conditions under which a licence for the use of the ISI Certification Mark may be granted to manufactures or processors, may be obtained from the Indian Standards Institution.

6 Sampling

6.1 The method for drawing representative samples and the criteria for conformity shall be as given in Appendix E of IS: 2236-1968[1].

---

[1] Specification for prawns/shrimp canned in brine (*first revision*).

7 Tests

7.1 Tests shall be carried out as prescribed in the relevant appendices in IS: 2168-1971[2] and IS: 2236-1968[1] as specified in column 4, 5 and 6 of Table 1 and column 4 of Table 2.

7.2 Quality of reagents – Unless specified otherwise, pure chemicals and distilled water (*see* IS: 1070-1960[3]) shall be used in tests.

> **Note** – 'Pure chemicals' shall mean chemicals that do not contain impurities which affect the test results.

[2] Specification for pomfret canned in oil (*first revision*).
[3] Specification for water, distilled quality (*revised*).

Appendix 3
Irish standard specification
181: 1971

FROZEN SMOKED DEMERSAL FISH FILLETS

Scope

1 This specification applies to whole fillets or large pieces of whole fillets cut from demersal fish (i.e. fish which swim close to the bottom of the sea) that have been boned, dyed (if desired), cold smoked, packaged, frozen and maintained at the temperature necessary for the preservation of the product.

Types of fillet

2 Fish may be 'single' fillets which are cut away from either side of the fish or 'block' fillets which incorporate both sides of the fish in one piece. Fillets may be either skin-on or skinless.

Freezing rate

3 The fillets shall be frozen at a rate of at least ¼ in (6 mm) per hour from both sides until the temperature at the centre of the product has reached −18°C and shall be continuously maintained at or below that temperature during storage.

Wholesomeness

4 The finished product shall be clean, wholesome and in every way fit for human consumption.

Definitions

5 (i) *'Slight'* refers to a condition that is scarcely noticeable but that does affect the appearance, desirability and/or eating quality of the fillets.

(ii) *'Moderate'* refers to a condition that is conspicuously noticeable, but that does not seriously affect the appearance, desirability and/or eating quality of the fillets.

(iii) *'Excessive'* refers to a condition that is conspicuously noticeable and that does seriously affect the appearance, desirability and/or eating quality of the fillets.

(vi) *'Cutting and trimming imperfections'* refers to fillets that have ragged edges, tears, or are otherwise improperly cut or trimmed so as to impair the visual desirability or usability of the product.

(v) *'Blemish'* means an instance as defined below of skin (for skinned fillets), scales, blood spot, bruise, black belly lining, fin or extraneous material.

(a) *'Instance of skin'* refers to a piece of skin not less than 1 cm^2 and not more than 2 cm^2 in area. Each additional area of up to 1 cm^2 of skin greater than 2 cm^2 shall be considered an additional instance.

(b) *'Instance of scales'* – each aggregate area of isolated scales or group of scales of up to 5 cm^2 per fillet or portion.

(c) *'Instance of blood spot'* refers to a spot of at least 3 mm in its maximum diameter. Each additional 3 mm in diameter shall be considered an additional instance.

(d) *'Instance of bruise'* refers to one of not less than 1 cm^2 and not more than 2 cm^2 in area. Each additional area of up to 2 cm^2 shall be considered an additional instance.

(e) *'Instance of black belly lining'* refers to any piece of black belly lining not less than 1 cm and not more than 2 cm in length. Each additional length of up to 1 cm of black belly lining longer than 2 cm shall be considered an additional instance.

(f) *'Instance of fin'* – each aggregate of identifiable fin, or part of any fin up to 5 cm^2 in area.

(g) *'Extraneous material'* – any piece of foreign matter on the product or elsewhere within the package.

(iv) *'Bone'* refers to a bone that after the product has been cooked is capable of piercing or hurting the palate.

Physical requirements

6 The product shall be examined in the frozen state for the factors of improper packing and desiccation (freezer burn) and in the thawed state for appearance, springiness, freedom from defects, size and odour. The product shall be cooked as described in Appendix A and examined for texture and flavour.

In these examinations the product shall be evaluated in accordance with the schedule in Table 1. For each factor other than odour and flavour the evaluation shall be by numerical scoring, points being deducted for variation of quality as indicated in the table. The total of the points deducted is subtracted from 100 to obtain

Table 1 Schedule of point deductions.

Scored factors	Description of quality variation		deduct points
Frozen products			
1 Improper packing	Imbedded packaging material, inadequate closure of water-vapour-proof material	Slight	1
		Moderate	2
		Excessive	3
2 Desiccation (freezer burn)	Slight, shallow and not colour masking, affecting an area of	5 to 50%	1
		over 50%	2
	Moderate – just deep enough to scrape off easily, affecting an area of	5 to 25%	3
		20 to 50%	8
		over 50%	16
	Excessive – deep dehydration not easily scraped off, affecting an area of	up to 5%	4
		over 5%	16
3 Appearance (colour)	A Flesh surface – loss of gloss, presence of white salt patches oily, gaping	Slight	2
		Moderate	4
		Excessive	8
	B Flesh – loss of characteristic yellow/brown colour and darkening of flesh	Slight	2
		Moderate	4
		Excessive	16
4 Springiness	Limp, soggy, dry when compared with freshly smoked flesh	Slight	2
		Moderate	4
		Excessive	8
Thawed products			
5 Freedom from defects	A Cutting and trimming imperfections affecting	Up to ¼ of total number of fillets per pack	8
		Over ¼ of total number of fillets per pack	16

Table 1 (Cont'd)

Scored factors	Description of quality variation		deduct points
	B Blemishes – number of instances per 500 g of fish flesh	1 or 2	2
		3	4
		3	4
		4	8
		5/6	12
		Over 6	16
	C Bones – number per 500 g of fish flesh	1	4
		2	8
		Over 6	16
	D Parasites (nematodes) – number per 500 g of fish flesh	1	1
		2	5
		Over 2	16
6 Size	Fillets or pieces less than 55 g	One fillet or piece or 10% of the number in the package, whichever is	
		the greater	1
		Exceeding the above	16
7 Odour	Fresh characteristic smoked fish odour	Category A	
	Absence of fresh, smoked odours – neutral odours	Category A	
	Smoked characteristics – acrid, resinous or lack of smoke	Category R	
	Slightly musty, milky, yeasty, oil, sour ammoniacal, putrid or faecal odours	Category R	
Cooked products			
8 Texture	Tough, dry fibrous or watery when compared with freshly smoked fish	Slight	4
		Moderate	8
		Excessive	16
9 Flavour	Full fresh flavour characteristic of freshly smoked fish	Category A	
	Slight sweetness and loss of characteristic flavour	Category A	
	Neutral flavour, definite loss of flavour as if chewing cotton wool	Category R	
	Sourness, some 'off' flavours, some bitterness some rancidity	Category R	
	Strong bitter flavours, slight sulphide-like flavours	Category R	
	Strong 'off' flavours of sulphides, putrid	Category R	

the score. Odour and flavour shall be classed in accordance with the schedule in Table 1.

The score of the product shall be not less than 85 points and the product shall be classed as Category A for both odour and flavour.

Package

7 Each package of frozen product shall contain only one species of fish. Each package shall weigh not less than 170 g (6 oz). Individual fillets or pieces shall not exceed 110 mm (4 in) in thickness. Fillets or pieces shall be wrapped or interleaved in such a manner that individual fillets can be separated without thawing.

The product shall be sealed in water-vapour-proof material, having a permeability of not more than 22.5 mg water vapour per 24 hours/m²/mm thickness/mbar at 25°C. The sealing material and all packaging material enclosed by it shall not contaminate the product or impart a flavour to it or in any way cause colour change of the product and shall not be discoloured by contact with the product. There shall be no printing ink on the surface of the packaging material in contact with the product.

Marking

8 The following information shall appear legibly and indelibly on each package or label:

(a) The name of the product as used by the trade in the country in which the product will be consumed and the type of fillet;
(b) the full name and business address of the manufacturer of the fillets; **or** in the case of packages for any trader who assumes full responsibility for them the full name and business address of that trader;
(c) the net weight of the contents in plain type of not less than 10 point size and in a colour which affords a distinct contrast with the colour of the package or label;
(d) the date of processing which may be in code;
(e) words indicating the country of origin;
(f) the inscription IS 175:1971; and

(g) directions for use which shall include the following statement: **'IF THAWED DO NOT REFREEZE'**

Sampling

9 *'Lot'* shall mean that quantity of frozen fillets from the same manufacturer of the same kind, type and package size submitted at any one time for inspection and testing to determine compliance with this specification.

'Defective' shall mean that package or its contents of frozen fish which does not comply in one or more respects with the requirements of this specification.

The following sampling procedure shall be applied:

Draw at random from the lot the relevant number of packages of frozen fillets given in Column 2 of Table 2. If the packages of frozen fillets are packed in large containers, draw at random the relevant number of containers, and from each draw at random one package. If the number of containers is less than the number of packages required, draw at least one package from each container.

The amount to be examined from each package shall be approximately 1.5 kg or the entire contents of the package, whichever is the less.

Compliance

10 The lot sampled in accordance with Clause 9 shall be deemed to comply with the requirements of this specification, if on inspection and testing of the sample for compliance with the requirements of this specification, not more than the relevant number of defectives indicated in Column 3 of Table 2 are found.

Table 2

1	2	3
Number of packages in lot	Number of packages in sample	Number of defectives (Acceptance No.)
Up to 1200	20	1
1201 to 10 000	32	2
10 001 to 35 000	50	3

APPENDIX A

Cooking for evaluation of flavour and texture

Approximately 100 g of the product shall be cooked from the frozen or thawed state, in a covered oven-proof glass container, in an atmosphere of steam, for a time appropriate to the thickness of the pieces(s).

Appendix 4
United States standards for grades of whole or dressed fish (Revised 1 October 1982)

Scope and product description

This standard shall apply to whole or dressed fish, whether fresh or frozen, of any species suitable for use as human food and processed and maintained in accordance with good manufacturing practices.

Product forms

(a) *Types:*
(1) Fresh.
(2) Frozen solid packs; glazed or unglazed.
(3) Frozen individually; glazed or unglazed.

(b) *Styles:*
(1) Whole.
(2) Dressed-eviscerated.
(3) Head-on or headless.
(4) With or without fins.
(5) Skin-on scaled or unscaled; semi-skinned (epidermis removed) or skinless.
(6) Other (as specified).

Grades – quality factors

(a) *US Grade A*. Whole or dressed fish shall:
(1) Possess good flavour and odour
(2) Comply with limits for defects for US Grade A quality in accordance with next section.

(b) *US Grade B.* Whole or dressed fish shall:

(1) Possess reasonably good flavour and odor and;

(2) Comply with the limits for defects for US Grade B quality in accordance with next section.

(c) *Substandard.* Whole or dressed fish does not possess reasonably good flavor and odor and/or exceeds the limits for defects for US Grade B quality in accordance with next section.

Determination of grade

(a) *Procedures for grade determination.*
The grade shall be determined by sampling in accordance with the sampling plan described in paragraph (b) of this section evaluating odour and flavour in accordance with paragraph (c) of this section examining for defects in accordance with paragraphs (d), (e) and (f) of this section and using the results to assign a grade as described in paragraph (g) of this section.

(b) *Sampling.*
The sampling rate of specific lots for all inspections, other than for military procurement, shall be in accordance with the sampling plans contained in Part 260 of this chapter except that the sampling unit is ten (10) fish for fish weighing up to 10 pounds. Fish weighing over ten (10) up to fifty (50) pounds – the sample unit shall be five (5) fish. For fish weighing over fifty (50) pounds, the sample unit shall be a minimum of three (3).

(c) *Evaluation of flavour and odour*

(1) Evaluation of the odour on each of the raw fish in the sample unit shall be carried out as follows:

(i) For this examination of small units, break the flesh or thawed sample either with the thumbs or by cutting with a knife in several places. Hold the cut or broken flesh close to the nose for evaluation.

(ii) For the examination of large units, a core may be used. Drill a hole into the hard frozen fish with a high-speed quarter inch drill. As soon as the drill is withdrawn, the hole and drillings are smelled.

(2) If the results of the raw odour evaluation indicate the existence

of any off-odours, the sample shall be cooked by any of the methods set forth below to verify the flavour and odour.

(i) *Boil in bag method*. Insert the sample into a boilable film-type pouch; fold the open end of the pouch over a suspension bar and clamp in place to provide a loose seal after evacuating the air by immersing the pouch into boiling water. Cook the contents for 20 minutes (until the internal temperature of the product reaches 160°F).

(ii) *Steam method*. Wrap the sample in a single layer of aluminum foil, and place on a wire rack suspended over boiling water in a covered container. Steam the packaged product for 20 minutes.

(iii) *Bake method*. Package the product as previously described. Place the packaged product on a flat cookie sheet or shallow flat-bottom pan of sufficient size so that the packages can be evenly spread on the sheet or pan. Place the pan and frozen contents in a properly ventilated oven preheated to 400°F for 20 minutes.

(3) The amount of material to be cooked shall be based on the results of the raw odour evaluation. A minimum of 25% of the sample except that not less than 3 sample units shall be used.

(d) *Examination for physical defects.*
Each of the fish in the sample will be examined for defects using the list of defect definitions, and the defects noted and categorised as minor, major, and serious in accordance with Table 1.

(e) *Definitions of defects in whole or dressed fish.*
(1) 'Abnormal condition' means that the normal physical and/or chemical structure of the fish flesh has been sufficiently changed so that the usability and/or desirability of the fish is adversely affected. It includes, but is not limited to, the following examples:

(i) Jellied – refers to the abnormal condition wherein a fish is partly or wholly characterised by a gelatinous, glossy, translucent appearance.

(ii) Milky – refers to the abnormal condition wherein a fish is partly or wholly characterised by a milky-white, excessively mush, pasty, or fluidised appearance.

(iii) Chalky – refers to an abnormal condition wherein a fish is partly or wholly characterised by a dry, chalky, granular appearance, and fibreless structure.

(A) Moderate – refers to a condition that is distinctly noticeable but does not seriously affect the appearance, desirability and/or the eating quality of the product.

Table 1

	Physical defects	Categories		
Types	Degree	Minor	Major	Serious
Abnormal	Moderate		201	
condition	Excessive			301
Appearance	Slight	102		
defects	Moderate		202	
	Excessive			302
Discoloration	Slight	103		
	Moderate		203	
	Excessive			303
Dehydration	Slight – more than 3% area affected and easily removed	104		
	Moderate – less than 3% area affected but difficult to remove.		204	
	Excessive – greater than 3% area affected.			304
Surface defects	Slight – 3–10% area affected	105		
	Moderate – greater than 10% area affected			305
Cutting and	Body cavity cuts	106		
trimming	Improper heading:			
defects	Slight	107		
	Moderate		206	
	Evisceration defects:			
	Slight	108		
	Moderate		207	
	Excessive			305
	Improper washing	109		
	Belly burn		208	
Texture	Slight	110		
defects	Moderate		209	
	Excessive			306

Note: The code numbers shown in the above table are for identification of defects for recording purposes only and are keyed to the nature and severity of the defect. They are not scores.

(B) Excessive – refers to a condition which is both distinctly noticeable and seriously objectionable.

(2) 'Appearance defects' shall refer to the overall general appearance of the fish (consistency of the flesh, odour, eyes, gills, and skin) and presence of excessive blood or drip and appearance of the package.

　(i) Slight – refers to an appearance defect that is slightly noticeable but does not seriously affect the appearance desirability, and/or eating quality of the fish.

　(ii) Moderate – refers to an appearance defect that is conspicuously noticeable but does not seriously affect the appearance, desirability, and/or eating quality of the fish.

　(iii) Excessive – refers to an appearance defect that is conspicuously noticeable and that does seriously affect the appearance, desirability and/or eating quality of the fish.

(3) 'Discoloration' refers to any colour not characteristic of the species used.

　(i) Slight – refers to the area affected by discoloration of significant intensity involving up to 10% of the total area.

　(ii) Moderate – refers to the area affected by discoloration of significant intensity involving over 10% and up to 50% of the total area.

　(iii) Excessive – refers to the area affected by discoloration of significant intensity involving over 50% of the total area.

(4) 'Dehydration' refers to loss of moisture from fish surfaces during frozen storage. For skin-on fish, dehydration shall be evaluated by degree of dullness and shrinkage.

　(i) Slight dehydration – is surface color masking affecting more than 3% of the area which can be readily removed by scraping with a blunt instrument.

　(ii) Moderate dehydration – is deep color masking penetrating the flesh, affecting less than 3% of the area, and requiring a knife or other sharp instrument to remove.

　(iii) Excessive dehydration – is deep color masking penetrating the flesh, affecting more than 3% of the area, and requiring a knife or other sharp instrument to remove.

(5) 'Surface defects' shall refer to the following where applicable:

(i) Scales. An occurrence of attached or loose scales in any sample unit (where applicable).

(ii) Blood spot. An accumulation of coagulated opaque, masses of blood on a fish.

(iii) Fins or pieces of fin. An occurrence or absence of attached or loose fins or pieces of fin in any sample unit (where applicable). Dorsal spine shall be removed (where applicable).

(iv) Skin. The presence of the dark or light inner layers of skin for skinless. For semi-skinned, reference is to the presence of the dark outside layers.

(v) Bruises. An accumulation of damaged portions of fish muscle, red and opaque in appearance (on a fish).

(vi) Damage to protective coating refers to voids in ice glaze or tears in covering membrane, also to breaks or splits in the skin which are readily discernible and not normally part of the processing.

(6) 'Cutting and trimming defects' refers to the following:

(i) Body cavity cuts – refers to misplaced cuts made during evisceration.

(ii) Improper heading (as specified) – refers to the presence of pieces of gills, gill cover, pectoral fins (spine), or collarbone after the fish have been headed. No ragged cuts should be evident after heading.

(iii) Evisceration defects – refers to inadequate cleaning of the belly cavity of the fish. All viscera, kidney (where applicable), spawn, and blood should be removed.

(A) Slight degree of improper evisceration and improper heading refers to a condition that is scarcely noticeable but does not affect the appearance, desirability, and/or eating quality of the fish.

(B) Moderate degree of improper evisceration and improper heading refers to a condition that is conspicuously noticeable but does not seriously affect the appearance, desirability, and/or eating quality of the fish.

(C) Excessive degree of improper evisceration refers to a condition that is conspicuously noticeable and that seriously affect the appearance, desirability, and/or eating quality of the fish.

(iv) Improper washing – inadequate removal of slime, blood, and bits of viscera from the surface of the fish and from the body cavity.

(v) Belly burn – an enzymatic action on the flesh causing a burned or discolored appearance.

(7) 'Texture defects' texture of the cooked fish; not characteristic of the species.

(i) Slight – fairly firm, only slightly tough or rubbery, does not form a fibrous mass in the mouth, moist but not mushy.

(ii) Moderate – moderately tough or rubbery, has noticeable tendency to form a fibrous mass in the mouth, moist but not mushy.

(iii) Excessive – excessively tough or rubbery, has marked tendency to form a fibrous mass in the mouth, or is very dry or very mushy.

(f) *Categorisation of physical defects. See Table 1*

(g) *Grade assignment.*

(1) Each fish in a sample unit will be assigned the grade into which it falls in accordance with the limits for defects, summarized as follows:

	Flavor and odor	Maximum number of physical defects permitted		
		Minor	Major	Serious
Grade A	Good	3	0	0
Grade B	Reasonably good	5	1	0

(2) Upon determination of grade of each fish in each sample unit, the sample will be designated a grade as follows:

(i) *Grade A.*

Number of subsample units (fish)	Minimum number of Grade A fish	Maximum number of Grade B fish	Maximum number of substandard
10 (up to 10 lb)	8	2	0
5 (10–50 lb)	4	1	0
3 (over 50 lb)	3	0	0

(ii) *Grade B.*

Number of subsample units (fish)	Minimum number of Grade B fish	Maximum number of substandard
10 (up to 10 lb)	8	2
5 (10–50 lb)	4	1
3 (over 50 lb)	3	0

(iii) *Substandard.* Any fish not meeting the minimum requirements for Grade B quality.

(3) Upon determination of the grade for each sample unit a lot of whole or dressed fish shall be assigned that grade in which:

(i) For physical defects, the number of sample units in the next lower grade does not exceed the acceptance number for deviants prescribed in Part 260.61 of the sampling plan, Table II, of Title 50; and

(ii) Not more than 5% of the fish in the sample (total fish examined per lot) are in the next lower grade for odour and/ or flavour.

Note – Sampling for inspection for military procurement shall be in accordance with MIL-STD-105. Lot size shall be expressed in terms of pounds. The sample size shall be in accordance with Inspection Level S-3. Acceptable Quality Levels shall be expressed in terms of defects per hundred units. The AQLs shall be 6.5 for minor and 4.0 for major.

Hygiene

Whole or dressed fish shall be processed and maintained in accordance with the applicable requirements of the regulations contained in §§ 260.96 to 260.103 of this chapter and of the good manufacturing practice regulations contained in 21 CFR Part 128.

Table 1 Schedule of point deductions per sample.

Factors scored	Method of determining score	Deduct
Frozen state (Lot inspection only)		
1 Arrangement of product[1]	Small degree: 10% of fish twisted or bellies and backs not facing the same direction.	2

Table 1 (Cont'd)

Factors scored	Method of determining score	Deduct
	Large degree: More than 10% of fish twisted, void present or some fish cross packed.	5
2 Condition of packaging (overall assessment)	Poor: Packaging material has been soaked, softened or deteriorated.	2
3 Dehydration	Small degree: Slight dehydration of the exposed surfaces	2
Thawed state		
4 Minimum size: Fish 2 oz or over are of acceptable size	Number of fish less than 2 oz per lb	
	Over 0 – not over 0.5	5
	Over 0.5 – not over 1.0	10
	Over 1.0 – not over 2.0	20
	Over 2.0	30
5 Uniformity: Weight ratio of fish remaining The 10% largest fish divided by the 10% smallest fish[1]	Weight ratio 10% smallest and 10% largest:	
	Over 2.0 – not over 2.4	2
	Over 2.4 – not over 2.8	5
	Over 2.8 – not over 3.2	10
	Over 3.2 – not over 3.6	20
	Over 3.6	30
6 Heading[1]	Small degree: 10% of fish carelessly cut	5
	Moderate degree: Over 10% of fish carelessly cut	15
7 Evisceration (overall assessment)	Small degree: Slight evidence of viscera	2
	Moderate degree: Moderate amounts of spawn, viscera, etc.	10
	Large degree: Large amounts of viscera, spawn, etc.	30
8 Scaling[1]	Small degree: 10% of fish not well scaled	2
	Large degree: Over 10% of fish not well scaled	5
9 Color of the exposed surfaces (overall assessment)	Small degree: Minor darkening, dulling	2
	Large degree: Objectionably dark, brown, dull	5
10 Bruises and split or broken skin	Presence of bruises and/or broken or split skin per pound:	
	Over 0 – not over 0.5	1
	Over 0.5 – not over 1.0	2
	Over 1.0 – not over 1.5	4
	Over 1.5 – not over 2.0	7
	Over 2.0	10
11 Texture: (overall assessment)	Small degree: Moderately dry, tough, mushy, rubbery, watery, stringy.	5
	Large degree: Excessively dry, tough, mushy, rubbery, watery, stringy.	15

[1] 10% of fish refers to 10% by count rounded to nearest whole fish.

Appendix 5

Recommended microbiological limits for seafoods (International Commission on Microbiological Specifications for Foods, 1986)

Product	Test	Limit[1] per g or per cm^2
Fresh, cold smoked and frozen fish	APC[2]	5.10^5
	E. coli	11
	Salmonella	0
	V. parahaemolyticus	10^2
	S. aureus	10^3
Precooked breaded fish	APC	5.10^5
	E. coli	11
	S. aureus	10^3
Frozen, raw Crustacea	APC	10^6
	E. coli	11
	Salmonella	0
	V. parahaemolyticus	10^2
	S. aureus	10^3
Frozen, cooked Crustacea	APC	5.10^5
	E. coli	11
	Salmonella	0
	V. parahaemolyticus	10^2
	S. aureus	10^3
Cooked, chilled and frozen crabmeat	APC	10^5
	E. coli	11
	V. parahaemolyticus	10^2
	S. aureus	10^3
Fresh or frozen bivalves	APC	5.10^5
	E. coli	16
	Salmonella	0
	V. parahaemolyticus	10^2

[1] Acceptable or attainable in good commercial practice
[2] Aerobic Plate Count (equivalent to Standard Plate Count)

Appendix 6
Excerpt from code of practice for frozen fish

Blast freezers should be loaded in such a way that there is a sufficient flow of cold air around the product

In this process, heat is transferred from the fish to a cold air stream and carried to the cooling surfaces of the freezer. Adequate air circulation is essential and any obstruction to the flow of air around the product will result in poor freezing rates and variable product quality. If fish are placed too close together because of overloading the freezer, cold air circulation around the surfaces of individual fish will be obstructed and freezing times may be greatly increased. Wrapping fish or placing it in cartons will also slow down the rate of freezing.

Large fish such as tuna, which are to be canned, should preferably be frozen by immersion in refrigerated brine

In order to minimise salt penetration and because it is impracticable to work with brine temperatures lower than $-18°C$ ($0°F$), fish frozen in this way should have their temperatures at the centre lowered as rapidly as possible to between $-12°C$ ($10°F$) and $-15°C$ ($5°F$). The temperature should then be lowered further to $-18°C$ ($0°F$) or below in storage. During freezing there should be a rapid circulation of the cooling medium to ensure effective heat transfer. An upward circulation will assist in keeping the fish in suspension and all their surfaces in contact with the cooling medium. To avoid unnecessarily high salt penetration, the fish should be either removed from the brine or the brine pumped out as soon as freezing is completed.

All freezing processes should be completed in the freezer by allowing the full time for each cycle

The manufacturer of the refrigeration equipment should provide all necessary information for the correct operation of the plant, including the time required for each freezing cycle. If the plant is functioning properly and loading and unloading is done according to instructions, fish coming out of the freezers should be properly frozen. There is always a temptation to reduce the freezing time, or overfill freezers during periods of heavy catching. This should be avoided. If the freezing time is too short, the centre of the block will not be frozen, even though the surface may be hard. Blocks of fish which are not completely frozen are easily broken during unloading and storing. If many partly frozen blocks are stored, the freezer store temperature may rise, placing an extra load on the refrigeration equipment and also causing temperature fluctuations that will adversely affect the quality of all the fish in storage.

On the other hand, if fish are left in the freezers long after they are properly frozen, freezer capacity is wasted and unnecessary delays in the freezing of fish will occur. In the case of air blast or sharp freezers, there will also be quality losses due to dehydration of the fish surfaces.

Fillets should be frozen rapidly to ensure a high quality product

Freezing of fillets should be carried out in contact or blast freezers. The use of brine is not recommended for the freezing of fillets because of salt penetration into the product.

Frequent checks should be made on the the pressures and temperatures in the refrigeration system to ensure correct operation

If frequent checks are made and records of these maintained, there will be little chance of the refrigerant's temperatures being too high or the equipment not functioning correctly. Any defects noted should be rectified quickly. It is important to watch the temperature gauges for superheating at the compressor's delivery side and sub-cooling of the liquid before the expansion valves. Sometimes, these two readings will indicate leaks of refrigerant before there is any serious loss of freezing capacity.

Accurate records of all freezing operations should be kept

An accurate record of all loading and unloading times of the freezer and number of blocks frozen, including size and species, will greatly assist in efficient management and control of the operations.

A system of labels or colour codes should be used when loading fish into a freezer to assist in the later identification of frozen products

Some system of identification is required to indicate the species, size, condition of fish and its suitability for further processing and handling.

Suggested further reading and reference

The ground covered by several of the following publications is very extensive and consequently their subject matter is relevant to more than one chapter of this book. Therefore any one publication is associated with the chapter that most nearly reflects its contents.

CHAPTER 2

Species – descriptions and illustrations

There is no single publication that includes descriptions and illustrations of the food fishes of all countries. The following is a list of the more useful books published in the last few decades dealing with the food fishes of individual countries or regions. Where books specifically on food fishes are not available, books are included that give information on all fish present in or near countries or regions; sometimes the commercial importance of the fish is stated.

The best world-wide checklist of food fishes is provided by the Yearbook of Fishery Statistics published annually by FAO.

American Fisheries Society (1970). *Special Publication No. 6. A List of Common and Scientific Names of Fishes from the US and Canada*, 3rd edition. Washington DC.

Bellisio, N.B., Copez, R.B. and Lorne, A. (1979). *Peces Marinos Patagonicas*. Ministerio de Economica, Buenos Aires.

Bini, G. (1965). *Catalogue of Names of Fishes, Molluscs and Crustaceans of Commercial Importance in the Mediterranean*. FAO and Vito Bianco Editore.

Blanc, M., Gaudet, J.-C., Banarescu, P. and Hurea, J.-C. (1971). *European Inland Water Fish*. FAO and Fishing News Books, Oxford, England.

Cabo, F.L., Martin, O.R. and Gratacos, P.A. (1965). *Nomenclatura*

Official Española de los Animales Marinos de Interes Pesquero. Direccion General de Pesca Maritima, Madrid.

Davidson, A. (1972). *Mediterranean Seafood.* Penguin Books, London.

Davidson, A. (1976). *Seafood of S.E. Asia.* Allan Davidson, London.

Davidson, A. (1979). *North Atlantic Seafood.* Macmillan, London.

Far Seas Fishery Research Laboratory (1972 and 1976). *Illustrations of Bottomfishes Collected by Japanese Trawlers.* Kanda, Tokyo.

Food and Agriculture Organization (1973, 1974 and 1978). *Species Identification Data Sheets For Fishery Purposes.* FAO, Rome.

Gousset, J. and Tixerant, G. (1969 and 1970). *Inspection des produits de la pêche*; *1. Identification des poisson, 2 Identification des poisson de mers, 3. Identification: poisson d'eau douce: crustaces: mollusques.* Information Techniques des Direction des Services Vétérinaires. Revue Trimestrielle, nos. 28, 29 and 32.

Hart, J.L. (1973). *Pacific Fishes of Canada.* Fisheries Research Board of Canada, Ottawa.

Kainuma, M. (1969). *A Handbook of the Poisonous Fishes and Shellfish.* Tokyo Press Co. Ltd, Tokyo.

Leim, A.H. and Scott, W.B. (1966). *Fisheries of the Atlantic Coast of Canada.* Fisheries Research Board of Canada, Ottawa.

New Zealand Fishing Industry Board. (1981). *Guidebook to New Zealand Commercial Fish Species.*

Organisation for Economic Co-operation and Development. (1978). *Multilingual Dictionary of Fish and Fish Products.* Fishing News Books, Oxford, England.

Palombi, M. and Santarelli, M. (1961). *Gli Animali Commestibili dei Mari D'Italia.* Editore Ulrico Hoepli, Milan.

Palumbo, M. (1971). *Biologia Marina e Technica della Pesca, Volume Primo.* Edizione Laurenziana, Naples.

Perlmutter, A. (1961). *Guide to Marine Fishes.* Bramhall House, New York.

Pownall, P. (1977). *Commercial Fishes of Australia.* Australian Government Publishing Service, Canberra.

Roughley, T.C. (1966). *Fish and Fisheries of Australia.* Angus and Robertson, Sidney.

Sancho, J.G. (1986). *Names and Descriptions of Principal Marine Fishes of Portugal.* Instituto Nacional de Investigacao das Pescas. Publicacaos avulsas No. 9, Lisbon.

Waterman, J.J. (1972). *Fish Names in the Common Market.* Torry

Research Station, Advisory Note No. 55. Her Majesty's Stationery Office, Edinburgh, Scotland.

Watkin, G. (1976). *British Food Fish*. The Worshipful Company of Fishmongers, London.

Wheeler, A. (1969). *The Fishes of the British Isles and N.W. Europe*. Macmillan, London.

Wray, T. (1979). *Commercial Fishes of Saudi Arabia*. Ministry of Agriculture and Water Resources, Kingdom of Saudi Arabia and Sea Fish Industry Authority, Edinburgh.

Young, H.-C. and Chen, T.-P. (1971). *Common Food Fishes of Taiwan*. Chinese-American Joint Commission on Rural Reconstruction, Taipei, Taiwan.

Condition and composition

Borgstrom, G. (ed.) (1961). Various chapters in *Fish as Food*, Vol. 1. Academic Press, New York and London.

Bykov, V.P. (1985). *Marine Fishes: Chemical Composition and Processing Properties*. A.A. Balkema, Rotterdam.

Heen, E. and Kreuzer, R. (ed) (1962). *Fish in Nutrition*. Fishing News Books, Oxford, England.

Love, R.M. (1970 and 1980). *The Chemical Biology of Fishes*, Vols. 1 and 2. Academic Press, London and New York.

Murray, J. and Burt, J.R. (1969). *The Composition of Fish*. Torry Research Station Advisory Note No. 38. Her Majesty's Stationery Office, Edinburgh, Scotland.

Sidwell, V.D. (1981). *Chemical and Nutritional Composition of Finfishes, Whales, Crustacea, Molluscs and Their Products*. National Oceanic and Atmospheric Agency, Technical Memo. F/SEC11. Seattle.

Parasites

Cheng, T.C. (1973). Human parasites transmissible by seafood. In *Microbial Safety of Fishery Products*, Chichester, C.O. and Graham, H.D. (ed.). Academic Press, New York and London.

Hoffman, G.L. and Sinderman, C.J. (1962). *Common Parasites of Fish*. Bureau of Commercial Fisheries Circular No. 144, US Department of the Interior.

Houwing, H. (1969). *The Inactivation of Herring Nematodes* (Anisakis marina) *by Freezing*. International Institute of Refrigeration, Bulletin Annex 1969–6, p. 297.

Olsen, R.E. (1987). *Marine Fish Parasites of Public Health Importance* in *Seafood Quality Determination*, Kramer, D.E. and Liston, J. (ed.). Elsevier, Amsterdam.

Rae, B.B. (1972). Parasites in fish. *Pisces*, **1** (4) p. 2.

Williams, H.H. and Jones, A. (1976). *Marine Helminths and Human Health*. Commonwealth Institute of Helminthology, Miscellaneous Publication No. 3, London.

von Bonsdorff, B. (1977). *Diphyllobothriasis in Man*. Academic Press, London and New York.

Toxic Fish

Bagnis, R. *et al.* (1970). Problems of toxicants in marine food. *Bull. World Health Organisation*, **42**, 69.

Halstead, B.W. (1965–70). *Poisonous and Venomous Marine Animals of the World*, Vol. 1–3. US Government Printing Office, Washington DC.

Ragelis, E.P. (1984). *Seafood Toxins*. American Chemical Society, Symposium Series 262, Washington.

Russel, F.E. (1969). Poisons and venoms. In *Fish Physiology*, Vol. III, Hoar, W.S. and Randall, D.J. (ed.), Academic Press, London and New York.

Scheuer, P.J. (1970). Toxins from fish and other marine organisms. *Advances in Food Research*, **18**, 141.

Pollution

Clarke, R.B. (ed.) (1982). *The Long-term Effects of Oil Pollution on Marine Populations, Communities and Ecosystems*. The Royal Society, London.

Goldberg, E.G. (ed.) (1972). *A Guide to Marine Pollution*. Gordon and Breach Science Publishers; New York, London and Paris.

Holden, A.V. (1973) Mercury in fish and shellfish. A review. *J. Fd Tech.* **8**, 1.

Johnston, R. (ed.) (1976). *Marine Pollution*. Academic Press, London and New York.

Motohiro, T. (1983). Tainted fish caused by petroleum compounds: a review. *Water Science and Technology*, **15**, 75.

Nauen, C.E. (1983). *Compilation of Legal Limits for Hazardous Substances in Fish and Fish Products*. FAO Fisheries Circular No. 764, Rome.

Newberne, P.M. (1973). Mercury in fish – a literature review. *CRC Critical Reviews in Food Technology*, **4**, (3), 311.

Ruivo, M. (ed) (1972). *Marine Pollution and Sea Life*. Fishing News Books, Oxford, England.

Tidmarsh, W.G. *et al.* (1985). *Tainting in Fishery Resources*. Department of Fisheries and Oceans, Ottawa.

Whittle, K.H. (1978). *Tainting in Marine Fish and Shellfish*. International Register of Potentially Toxic Chemicals, Vol II, pp. 89–108. UN Environment Program, Geneva.

Occasional peculiarities

Ackman, R.G. *et al.* (1967). Dimethyl-*B*-propiothetin and dimethyl sulphide in Labrador cod, *J. Fish. Res. Bd Can.* **24**, 457.

Ackman, R.G. *et al.* (1972). Dimethyl sulphide as an odour component in Nova Scotia fall mackerel. *J. Fish Res. Bd Can.* **29**, 1085.

Maudesley-Thomas, L.E. (ed.) (1972). *Diseases of Fish*. Academic Press, London and New York.

Persson, P.-E. (1981). The etiology of muddy odour in water and fish. *Finnish Fisheries Research*, **4**, 1.

Sindermann, C.J. (1970). *Principal Diseases of Marine Fish and Shellfish*. Academic Press, London and New York.

Snieszko, S.F. (ed.) (1970). *A Symposium on Disease of Fish and Shellfish*. Special Publications No. 5. American Fisheries Society.

CHAPTERS 3, 4 AND 5

Aitken, A. *et al.* (ed.) (1982). *Fish Handling and Processing*, 2nd edition. Her Majesty's Stationery Office, Edinburgh.

Berka, R. (1986). *Transport of Live Fish: a Review*. EIFA Technical Paper No. 48, FAO, Rome.

Bostock, T.W., Walker, D.J. and Wood, C.D. (1987). *Reduction of Losses in Cured Fish in the Tropics*. Tropical Development and Research Institute, London.

Borgstrom, G. (ed.) (1961–65). Various chapters in *Fish as Food*. Vol. I–IV. Academic Press, New York and London.

Bramsnaes, F. (1969). Quality and stability in frozen sea-food. In *Quality and Stability in Frozen Foods*, Van Arsdel, W.B., Copley, M.J. and Olson, R.L. (eds). Wiley-Interscience, New York.

Chichester, C.O. and Graham, H.D. (ed.) (1973). *Microbial Safety of Fishery Products*. Academic Press, New York and London.

Dore, I. (1984). *Fresh Seafood: The Commercial Buyer's Guide*. Osprey Books, Huntington.

Doyle, J.P. (1974). *Fishplant Sanitation and Cleaning Procedures*. Marine Advisory Bulletin No. 1, University of Alaska.

Food and Agriculture Organisation/International Atomic Energy/World Health Organisation. Expert Committee. (1981). *Wholesomeness of Irradiated Food*. WHO Technical Report Series No. 659. Her Majesty's Stationery Office, London.

Gousset, J. and Tixerent, G. (1971). *Inspection des Produits de la Pêche 4*. Fraicheur et altération. Techniques de l'inspection conseils d'hygiene. Information techniques des Directions des Services Vétérinaires, No. 34.

Hennig, R. (1972). *Fischwaren*. VEB Fachbuchverlag, Leipzig.

Kietzmann, V., Priebe, K., Rakow, D. and Reichstein, K. (1969). *Seefisch als Lebensmittel*. Paul Parey, Berlin and Hamburg.

Kreuzer, R. (ed.) (1969). *Freezing and Irradiation of Fish*. Fishing News Books, Oxford, England.

Lane, J.P. (1974). *Sanitation Recommendations for Fresh and Frozen Fish Plants*. Fishery Facts No. 8, National Marine Fisheries Service, Seattle, Washington, D.C.

Ludorff, W. and Meyer, V. (1973). *Fische und Fischzeugnisse*, 2nd edition. Verlag Paul Parey, Berlin and Hamburg.

Maestrelli, A. and Senesi, E. (1984). *Consumare il Mare*. Clesav, Milan.

National Canners Association (1973). *Principles of Thermal Process Control and Container Closure Evaluation*. USA.

Nichelson II, R. (1972). *Seafood Quality Control – Boats and Fish Houses: Processing Plant*. Marine Advisory Bulletins, Texas A and M University, Texas.

Nickerson, J.T. and Sinskey, A.J. (1972). *Microbiology of Foods and Food Processing*. American Elsevier Publishing Co., New York.

Nokikov, V.M. (1983 and 1984). *Handbook of Fishery Technology*.

Vol. 1 and 4. A.A. Bulkema, Rotterdam.

Reilly, A. (ed.) (1985). *Spoilage of Tropical Fish.* FAO Fisheries Report No. 317, Rome.

Pesle, O. and Schwierrzina, A. (1969). *Qualität von Fisch und Fischwaren.* VEB Fachbuchverlag, Leipzig.

Reilly, A. and Barile, L.E. (ed.) (1986). *Cured Fish Production in the Tropics.* University of the Philippines, Visayas.

Sainclivier, M. (1983 and 1986). *L'Industrie Alimentaire Halientique, Vol. 1 and 2.* Science Agronomiques, Rennes.

Sikorski, Z.E. (1980). *Technologia Zywnosii Pochodzenia Marskiego.* Wydawnictwa Nankowo-Techniczne, Warsaw.

Stumbo, C.R. (1973). *Thermobiology in Food Processing,* 2nd edition. Academic Press, New York and London.

Tanikawa, E. (1971). *Marine Products in Japan.* Koseisha-Koseikaku Co., Tokyo.

Various Authors (1960 – present) *Advisory Notes* – Torry Research Station, Her Majesty's Stationery Office, Edinburgh.

Walker, D.J. (1987). *Review of Use of Contact Insecticides.* FAO Fisheries Circular No. 804, Rome.

Wheaton, F.W. and Lawson, T.B. (1985). *Processing Aquatic Food Products.* Wiley, New York.

CHAPTER 6

Connell, J.J. and Howgate, P.F. (1986). *Fish and Fish Products* in *Quality Control in the Food Industry,* 2nd edition, Vol. 4, Hoerschdoerfer, S.M. (ed.). Academic Press, London.

Durand, P., Landrein, A. and Quero, J.C. (1985). *Electrophoretic Catalogue of Commercial Fish Species.* Institut Francais de Recherche pour l'Exploitation de la Mer, Nantes.

Farber, L. (1965). Freshness tests. In *Fish as Food,* Borgstrom, G. (ed.), Vol. IV, p. 65. Academic Press, New York and London.

Gould, E. and Peters, J.A. (1971). *On Testing the Freshness of Frozen Fish.* Fishing News Books, Oxford, England.

Hennings, C. (1965). The 'Intelectron Fish Tester IV' – a new electronic method and device for rapid measurement of the degree of 'wet' fish. In *The Technology of Fish Utilization,* Kreuzer, R. (ed.). Fishing News Books, Oxford, England.

International Commission on Microbiological Specifications for Foods. (1986). *Microorganisms in Foods – 2. Sampling for Microbiological Analysis.* 2nd edition. University of Toronto Press, Toronto.

Jason, A.C. and Lees, A. (1971) *Estimation of Fish Freshness by Dielectric Measurement*. Department of Trade and Industry, London.

Kramer, A. and Twigg, B.A. (1970). *Quality Control for the Food Industry*, 3rd edition, Vol. 1, *Fundamentals*. The Avi Publishing Co. Inc. Westport, Connecticut.

Kramer, D.E. and Liston, J. (ed.) (1987). *Seafood Quality Determination*. Elsevier, Amsterdam.

Laird, W.M., Mackie, I.M. and Ritchie, A.M. (1982). Isoelectric focussing in the identification of fish species. *J. Assoc. Publ. Anal.*, **20**, 125.

Lundstrom, R.C. (1981). Isoelectric focussing. *J. Assoc. Off. Anal. Chem.*, **64**, 38.

Piggott, J.R. (ed.) (1984). *Sensory Analysis of Foods*. Elsevier, Amsterdam.

Steiner, E.H. (1985). Statistical methods in quality control. In *Quality Control in the Food Industry*, 2nd edition, Vol. 1, Hoerschdoerfer, S.M. (ed.). Academic Press, London and New York.

Sutherland, J.P., Varnam, A.H. and Evans, M.G. (1986). *A Colour Atlas of Food Quality Control*. Wolfe Publishing Ltd, London.

Stone, H. and Sidel, J.L. (1985). *Sensory Evaluation Practices*. Academic Press, New York.

Williams, S. (ed.) (1984). *AOAC Manual*, Association of Official Analytical Chemists, Washington.

Woyeda, A.D., Shaw, S.J., Ke, P.J. and Burns, B.G. (1986). *Recommended Methods for Assessment of Fish Quality*. Canadian Technical Report of Fisheries and Aquatic Sciences No. 1448, Halifax, Canada.

CHAPTER 7

Couden, H.N. (1969). *Quality Management in Quality and Stability in Frozen Foods*, Van Arsdel, W.B., Copley, M.J. and Olson, R.L. (eds). Wiley-Interscience, New York.

Food and Agriculture Organization (1971). *Fish Inspection Programmes*, FAO Fishery Reports No. 114. FAO, Rome.

Hawthorne, J. (1985) The organization of quality control. In *Quality Control in the Food Industry*, 2nd edition, Vol. 1., Hoerschdoerfer, S.M. (ed.). Academic Press, London and New York.

Howgate, P.F. (1984). *Report on Quality Control and Inspection –*

Systems for Fish Products in INFOFISH Member Countries. INFOFISH, Kuala Lumpur.

Iles, A.R. (1980). Fish and fish products. In *Food Control in Action*, Dennis, P.O. *et al.* (eds). Applied Science Publishers, London.

Kreuzer, R. (ed.) (1971). *Fish Inspection and Quality Control.* Fishing News Books, Oxford, England.

Paquette, G.N. (1983). *Fish Quality Improvement: a Manual for Plant Operators.* Osprey Books, Huntington, New York.

Tropical Products Institute (1977). *Proceedings of the Conference on the Handling, Processing and Marketing of Tropical Fish.* Tropical Research and Development Institute, London.

CHAPTERS 8 AND 9

Blaufart, G.P. and Johnston, E.C. (1987). Voluntary US standards for grades of fishery products. In *Seafood Quality Determination*, Kramer, D.E. and Liston, J. (eds). Elsevier, Amsterdam.

Hobbs, G. (1970). Quality control and standards. *Proceedings, Institute of Food Science and Technology*, **3**, 98.

Howgate, P.F. (1987). Fish inspection and quality control in Europe. In *Seafood Quality Determination*, Kramer, D.E. and Liston, J. (eds). Elsevier, Amsterdam.

Hutchinson, J.M. (1987). The impact, acceptance and use of Codex international standards and codes of practice for fish and fishery products. In *Seafood Quality Determination*. Kramer, D.E. and Liston, J. (eds). Elsevier, Amsterdam.

International Institute of Refrigeration. (1979). *Recommendations for the Chilled Storage of Perishable Produce*, IIR, Paris.

International Institute of Refrigeration. (1986). *Recommendations for the Processing and Handling of Frozen Foods*, 3rd edition. IIR, Paris.

Sea Fish Industry Authority. (1984). *Specifications for the Purchase of Fish.* SFIA, Edinburgh.

Sea Fish Industry Authority. (1985). *Guidelines for the Handling of Fish Packed in Controlled Atmospheres.* SFIA, Edinburgh.

Sea Fish Industry Authority (1987). *Guidelines for the Handling of Chilled Fish by Retailers.* SFIA, Edinburgh.

Shewan, J.M. (1970). Bacteriological standards for fish and fishery products. *Chemistry and Industry* Feb. 7, 1970, p. 193.

Stutts, M.A., Gillespie, S.M. and Schwartz, W.D. (1974). *Operations Manual for Seafood Retailers.* Texas Parks and Wildlife Department, Austin, Texas.

New Zealand Fishing Industry Board. (1981). *Code of Practice for Eel Processing*. NZFIB, Wellington.

New Zealand Fishing Industry Board. (1982). *Code of Practice for Mussels*. NZFIB, Wellington.

New Zealand Fishing Industry Board (1982). *Code of Practice for Air Freight of Chilled Fish*. NZFIB, Wellington.

Index

Books published by
Fishing News Books

Free catalogue available on request from Fishing News Books, Blackwell Scientific Publications Ltd, Osney Mead, Oxford OX2 0EL, England

Advances in fish science and technology
Aquaculture in Taiwan
Aquaculture: principles and practice
Aquaculture training manual
Aquatic weed control
Atlantic salmon: its future
Better angling with simple science
British freshwater fishes
Business management in fisheries and
  aquaculture
Cage aquaculture
Calculations for fishing gear designs
Carp farming
Commercial fishing methods
Control of fish quality
Crab and lobster fishing
The crayfish
Culture of bivalve molluscs
Design of small fishing vessels
Developments in electric fishing
Developments in fisheries research in
  Scotland
Echo sounding and sonar for fishing
The economics of salmon aquaculture
The edible crab and its fishery in British
  waters
Eel culture
Engineering, economics and fisheries
  management
European inland water fish: a multilingual
  catalogue
FAO catalogue of fishing gear designs
FAO catalogue of small scale fishing gear
Fibre ropes for fishing gear
Fish and shellfish farming in coastal waters
Fish catching methods of the world
Fisheries oceanography and ecology
Fisheries of Australia
Fisheries sonar
Fisherman's workbook
Fishermen's handbook
Fishery development experiences
Fishing and stock fluctuations
Fishing boats and their equipment
Fishing boats of the world 1
Fishing boats of the world 2
Fishing boats of the world 3
The fishing cadet's handbook
Fishing ports and markets
Fishing with electricity
Fishing with light

Freezing and irradiation of fish
Freshwater fisheries management
Glossary of UK fishing gear terms
Handbook of trout and salmon diseases
A history of marine fish culture in Europe and
  North America
How to make and set nets
Inland aquaculture development handbook
Intensive fish farming
Introduction to fishery by-products
The law of aquaculture: the law relating to the
  farming of fish and shellfish in Great Britain
The lemon sole
A living from lobsters
The mackerel
Making and managing a trout lake
Managerial effectiveness in fisheries and
  aquaculture
Marine fisheries ecosystem
Marine pollution and sea life
Marketing in fisheries and aquaculture
Mending of fishing nets
Modern deep sea trawling gear
More Scottish fishing craft
Multilingual dictionary of fish and fish
  products
Navigation primer for fishermen
Net work exercises
Netting materials for fishing gear
Ocean forum
Pair trawling and pair seining
Pelagic and semi-pelagic trawling gear
Penaeid shrimps — their biology and
  management
Planning of aquaculture development
Refrigeration of fishing vessels
Salmon and trout farming in Norway
Salmon farming handbook
Scallop and queen fisheries in the British Isles
Seine fishing
Squid jigging from small boats
Stability and trim of fishing vessels and other
  small ships
Study of the sea
Textbook of fish culture
Training fishermen at sea
Trends in fish utilization
Trout farming handbook
Trout farming manual
Tuna fishing with pole and line